# The Acoustics of the Social on Page and Screen

# The Acoustics of the Social on Page and Screen

Edited by
Nathalie Aghoro

BLOOMSBURY ACADEMIC
NEW YORK • LONDON • OXFORD • NEW DELHI • SYDNEY

BLOOMSBURY ACADEMIC
Bloomsbury Publishing Inc
1385 Broadway, New York, NY 10018, USA
50 Bedford Square, London, WC1B 3DP, UK
29 Earlsfort Terrace, Dublin 2, Ireland

BLOOMSBURY, BLOOMSBURY ACADEMIC and the Diana logo are trademarks of Bloomsbury Publishing Plc

First published in the United States of America 2022
This paperback edition published 2023

Copyright © Nathalie Aghoro, 2022
Each chapter copyright © by the contributor, 2022

For legal purposes the Acknowledgments on p. vi constitute an extension of this copyright page.

Cover design: Louise Dugdale
Cover image: Central Park/Orbon Alija/Getty Images

All rights reserved. No part of this publication may be reproduced or transmitted in any form or by any means, electronic or mechanical, including photocopying, recording, or any information storage or retrieval system, without prior permission in writing from the publishers.

Bloomsbury Publishing Inc does not have any control over, or responsibility for, any third-party websites referred to or in this book. All internet addresses given in this book were correct at the time of going to press. The author and publisher regret any inconvenience caused if addresses have changed or sites have ceased to exist, but can accept no responsibility for any such changes.

Whilst every effort has been made to locate copyright holders the publishers would be grateful to hear from any person(s) not here acknowledged.

Library of Congress Cataloging-in-Publication Data
Names: Aghoro, Nathalie, editor.
Title: The acoustics of the social on page and screen / edited by Nathalie Aghoro.
Description: New York : Bloomsbury Academic, 2021. | Includes bibliographical references and index. | Summary: "Sound positions individuals as social subjects. The presence of human beings, animals, objects, or technologies reverberates into the spaces we inhabit and produces distinct soundscapes that render social practices, group associations, and socio-cultural tensions audible. The Acoustics of the Social on Page and Screen unites interdisciplinary perspectives on the social dimensions of sound in audiovisual and literary environments. The essays in the collection discuss soundtracks for shared values, group membership, and collective agency, and engage with the subversive functions of sound and sonic forms of resistance in American literature, film, and TV"– Provided by publisher.
Identifiers: LCCN 2021013250 (print) | LCCN 2021013251 (ebook) | ISBN 9781501361388 (hardback) | ISBN 9781501361395 (epub) | ISBN 9781501361401 (pdf) | ISBN 9781501361418
Subjects: LCSH: United States–Civilization–1970– | Sound–Social aspects–United States. | Mass media and culture–United States. | Sound in literature | Sound in mass media.
Classification: LCC E169.12 .A215 2021 (print) | LCC E169.12 (ebook) | DDC 306.0973–dc23
LC record available at https://lccn.loc.gov/2021013250
LC ebook record available at https://lccn.loc.gov/2021013251

ISBN: HB: 978-1-5013-6138-8
PB: 978-1-5013-8341-0
ePDF: 978-1-5013-6140-1
eBook: 978-1-5013-6139-5

Typeset by Deanta Global Publishing Services, Chennai, India

To find out more about our authors and books visit www.bloomsbury.com and sign up for our newsletters.

# Contents

Acknowledgments ... vi

Introduction to *The Acoustics of the Social on Page and Screen*  Nathalie Aghoro ... 1

Part I  Sound Practice across Media

1  Listening in Print  *Nicole Brittingham Furlonge* ... 13
2  When a Poem "Sounds" through the Body  *Irene Polimante* ... 28
3  Practices of Unmixing: Film Aesthetics, Sound, and the New Hollywood Cinema  *Christof Decker* ... 44

Part II  Soundtracks of Collective Memory

4  Voice and Wake: Susan Howe, M. NourbeSe Philip, and the Ecology of Echology  *Julius Greve* ... 61
5  Reframing Indigenous Sonic Archives: Jeremy Dutcher and the Cultural Politics of Refusal  *Sabine Kim* ... 78
6  Unsettled Scores: Listening to Black Oklahoma on the American "Frontier"  *Tsitsi Jaji* ... 99

Part III  Social Acoustics and Politics of Sound

7  The Operator and the Final Girl: Gender, Genre, and Black Sonic Labor in *The Call*  *Allison Whitney* ... 115
8  Bohemian Like You: The Construction of Cool Sound Collectives in Serial Television  *Florian Groß* ... 130
9  Sonic Sites of Subversion: Listening and the Politics of Place in Karen Tei Yamashita's *Tropic of Orange*  *Nathalie Aghoro* ... 147
10 From "Dead Spots" to "Hot Spots": Ann Petry's "On Saturday the Siren Sounds at Noon"  *Jennifer Lynn Stoever* ... 164

Notes on Contributors ... 197
Index ... 200

# Acknowledgments

Brittingham Furlonge, Nicole. The contributed chapter was adapted from portions of *Race Sounds: The Art of Listening in African American Literature* by Nicole Brittingham Furlonge (Introductions and Chapter 3) © 2018 University of Iowa Press. Used with permission of the University of Iowa Press.

Civil, Gabrielle. Excerpts of *Swallow the Fish* © 2017 and *Experiment in Joy* © 2019 by Gabrielle Civil included with permission from the author and The Accomplices.

Frederick, Rico. Excerpts of *Broken Calypsonian* © 2014 by Rico Frederick. Published by Penmanship Books and reprinted with permission.

Frederick, Rico. Excerpt of "Heartbreaker" © 2011. Directed by Aron Baxter, produced by Nice Shoes, and included with permission from the author.

Howe, Susan. Excerpts of *Singularities* © 1990 by Susan Howe. Published by Wesleyan University Press and reprinted with permission.

Susan Howe. Excerpts of *STRAY: A GRAPHIC TONE* © 2019 by Shannon Ebner, Susan Howe, and Nathaniel Mackey. Published by Fonograf Editions / ROMA Publications and reprinted with permission.

Philip, M. NourbeSe. Excerpt of *She Tries Her Tongue, Her Silence Softly Breaks* © 1989 by M. NourbeSe Philip. Published by Wesleyan University Press and reprinted with permission.

Philip, M. NourbeSe. Excerpts of *Zong!* © 2008 by M. NourbeSe Philip. Published by Wesleyan University Press and reprinted with permission.

# Introduction to *The Acoustics of the Social on Page and Screen*

Nathalie Aghoro

Sound positions us as social participants. The presence of human beings, animals, objects, or technologies reverberates into the material places and virtual spaces we inhabit and produces distinct soundscapes that render social practices, group associations, and sociocultural tensions audible. On an everyday basis, the soundscapes in which we are embedded are a reminder that social life is perpetually in the making. In this light, they are generative because, as they mediate existing social configurations, they also invite participation.[1] Sound is a catalyst for interactions ranging from agreement to conflict and all the possible stages in between when we position ourselves in relation to the sounds we perceive and produce. The mere assessment of a sonic phenomenon as either noise or music, as rallying cry or public offense, as reassuring or disturbing event, already reveals the social and ideological paradigms that govern our perception.[2] As Michael Bull and Les Back write in *The Auditory Culture Reader*, paying critical attention to "sound makes us re-think . . . how we relate to others, ourselves, and the spaces and places we inhabit,"[3] and they argue that the study of sound has the capacity to reveal interconnections between social, cultural, economic, and political circumstances that would otherwise remain invisible or opaque.

*The Acoustics of the Social on Page and Screen* picks up on the observation that sound becomes a repository for the relational activities defining a living environment as it travels through space and unfolds its own temporality that is both transient and resonant. This edited collection of essays focuses on the cultural engagement with sound's social potential by addressing the intersubjective, participatory, and collective characteristics of sonic cultural production as well as the transversal dynamics of acoustic practices apt to challenge hegemonial power, question established social order and structures, and foster social change. Brandon LaBelle considers the soundscapes in which sound practice takes place to be acoustic territories, because changes in sonic practice, transgressions, or new additions to a soundscape are able to influence political dynamics in social collectives and provoke public processes of renegotiation in society at large. He argues that "[a]coustic space . . . brings forward a process of acoustic territorialization, in which the disintegration and reconfiguration of space . . . becomes a political process."[4] By linking acoustics to the notion of territory, LaBelle highlights the potential of sound to not only accompany negotiations of power within communities and reconfigurations of social order, but to actively initiate and promote them.

Studying the intersections of the sonic and the social through the lens of auditory cultures therefore means to add a significant dimension to the critical assessment of social collectives, their stratifications, constitutive processes, and their cultural representations. Kara Keeling and Josh Kun take the social and cultural specificities of sound into account, when they make a case for "listening to American Studies" in *Sound Clash*. In their introduction to this 2011 special edition of *American Quarterly*, they suggest a concerted scholarly effort to address the functions and features of the acoustics as an integral part of American culture, society, and politics from the role of sound in "debates around empire, immigration, and national culture," "citizenship," or "identities in the age of globalization."[5] Their advocacy for listening to the acoustics of social formations suggests a substantial link between the sounds of culture and social imaginaries like the nation, defined by Benedict Anderson as an "imagined political community,"[6] but also global collective configurations that transcend the nation like diasporas, social networks, subcultures, and other alternative forms of group alliance. The notion of "social imaginaries" is understood here in Charles Taylor's sense of the term—building on Anderson's notion of "imagined communities" and seeking to account for the multiple configurations of modernities around the globe.[7] Taylor writes that "[t]he social imaginary is not a set of ideas; rather, it is what enables, through making sense of, the practices of a society."[8] The emphasis on social practice as a constitutive element of social imaginaries foregrounds "the situated nature and collective forms of social interaction."[9] Sound practices and sonic representations directly engage with these two characteristics of the social since sound is inherently situated in the in-between—not least between the sites of enunciation and listening—with sound waves spreading through place and moving in between bodies. In this sense, the acoustics of the social reside in a field of tension between individual and collective.

Audiovisual, acoustic, and literary media can be considered as cultural sites where the capacity of sound to register and expose the stratifications and processes of society is negotiated—particularly when it comes to issues of access and participation, collaborative efforts, or the articulation of dissent and resistance. Sonic imaginaries are involved in the emergence and negotiation of relational dynamics, and they provide soundtracks for the expression of personal agency, shared values, and group membership. Sound practice across media takes part in perpetuating, reactivating, challenging, and transforming the fabric of the social and thus actively weighs in on alternative forms of coexistence and social futures. By listening critically to the acoustics of the social, this edited collection seeks to contribute to the conceptualization of acoustic literacy and sonic aesthetics in artistic practice and media representation. It aims at understanding the work sound does in the context of social imaginaries as they are received, challenged, forged, and projected on and beyond page and screen.[10] Two sonic modes of action are particularly relevant for cultural works that invite their audiences to hear, view, and read the social through the sounds it produces and resonates with. First, sound can be used to actively express the dynamics of power, cohesion, and change in social imaginaries. It is capable of both accentuating and contesting the appearance of relational constellations, as LaBelle observes: "[S]ound works to unsettle and exceed arenas of visibility by relating us to the unseen, the non-represented or the not-yet apparent; alongside spaces of appearance, and the legible visibilities often defining open discourse, the flow and force of particular tonalities and musics, silences and noises may transgress certain partitions of borders, expanding the agentive possibilities of the uncounted and the unheard."[11] Along these lines, sound can gesture to the past by reviving the history of represented inequities and to the future by carving out shifts and movements of transformation as they announce themselves. The expressive work of sound thus provides a critical gateway into the multidimensionality of social developments.

However, the work sound does in social contexts is contingent on the ideological orientations and regimes of power that determine which sound practices are developed, how they are carried out, and to what end or effect. This contingency leads to a second way in which sound works in social imaginaries: Sound puts normative forces into relief. Robin James calls the potential of sound to reinforce dominant discourses in society and police deviance from the norm the "sonic episteme," which she understands as "the neoliberal episteme's complementary qualitative episteme."[12] James argues that "[t]he sonic episteme creates qualitative versions of the same relationships that the neoliberal episteme crafts quantitatively, bringing nonquantitative phenomena in line with the same upgrades to classical liberalism that the neoliberal episteme performs quantitatively."[13] For instance, the sonic episteme provides a qualitative dimension to the structure of daily routines and behavioral norms in terms of pace and rhythm that James aligns with Lauren Berlant's take on neoliberalism as "establish[ing] norms governing the rate or pace of people's daily habits."[14] Coordinating with normative sounds synchronizes one's habits with "the rhythms of large-scale social institutions."[15] Thinking with sound therefore demands critical attention to the underlying assumptions of knowledge production. James cautions that "[s]ound . . . can be a productive model for theorizing if and only if it models intellectual and social practices that are designed to avoid and/or oppose the systemic relations of domination that classical liberalism and neoliberalism create."[16] Accordingly, her work engages with "the way people oppressed by those systems of domination think about and use sound"[17] to debunk normalizing discourses that determine social interaction and behavior. Alexander Weheliye's "phonographies" provide a framework for understanding both dominant sonic epistemes and acoustic counterdiscourses through his focus "on the conditions of im/possibility . . . [sound recording and reproduction] technologies engender for twentieth-century black culture and how black cultural sensibilities have shaped the history of sonic technologies."[18] Taking into account that sound—in particular, recorded, reproduced, and remediated sound—works in relation to technology, media, and other senses is an important premise for *The Acoustics of the Social*.

An overlap in scholarly concerns is noticeable throughout various kinds of sonic, audiovisual and textual media regarding the sonification of social imaginaries: From reflections on the orchestration of sound, its narrative functions, or the fusion of sonic aesthetics with image and text to ethical questions about the distribution of listening attention,[19] or the engagement with the transgressive and subversive functions of sound such as sonic forms of resistance against hegemonial practices of othering, gendering, and racializing sounding bodies.[20] When considering the intermedial constellations of sound with respect to audiovisual media and literature, the following questions arise: How does the aesthetic engagement with the sonic dimension across media incorporate the disruptive potential of sound into the imaginary worlds that these works project? Which sound collectives do films, literary works, music, and art register and how are they represented or mediated? And what are the societal and political functions of sound, noise, and music?

The cross-media approach to sound as repository for the social in various forms of cultural expression is fundamental to this edited collection and its three-part structure. The sections "Sound Practice Across Media," "Soundtracks of Collective Memory," and "Social Acoustics and Politics of Sound" all take into account sound, vision, and text in order to carve out points of convergence for in-depth understandings of sound aesthetics and acoustic literacy and their significance for the exploration of social issues in cultural imagination. Each of the essays analyzes sound in specific media, ranging from literature on and beyond the page—such as novels, poetry, oral literature, and song—to music and oral performance, as well as sound practice in TV series and film.

Part I, "Sound Practice Across Media," begins with critical reflections on the interconnections between sound and text, sound and vision, as well as sound and the representation of social dynamics. The focus on the media contexts in which practices such as listening, voicing, and sound mixing are deployed positions them as literary, cinematic, and performative devices suited to explore the driving forces of society continuously in the making. These practices, as the first three chapters show, are used and reproduced to relate the individual to a collective, to make social stratifications and negotiation processes audible, and to connect individual expression to the sonority of societal action.

In "Listening in Print," **Nicole Brittingham Furlonge** demonstrates that listening opens up productive pathways into literary forms of signification. She explores how novels by Toni Morrison, Nathaniel Mackey, and Gayl Jones prompt readers to listen and "tune in as a multisensory being to the sounds embedded in print materials." In the process, she elaborates on the concept of "listening in print" that she coined in her monograph *Race Sounds: The Art of Listening in African American Literature*.[21] Her attuned reading of the resonant mutuality of speech and sound and its relevance for social transformation in works like *Jazz* and *Corregidora* draws special attention to the democratic and ethical dimensions of listening as a literary sound practice. **Irene Polimante**'s "When a Poem 'Sounds' through the Body" turns toward the layered mediality of poetry from text to oral, corporeal, and audiovisual performance. Her investigation of Black twenty-first-century poetry brings together Rico Frederick's multimedia experiments with the distinct sonorities of Creole and English vernaculars, Gabrielle Civil's performative explorations of the physicality and connectivity of voice, and Terrance Hayes's play with form, language, and affect giving rise to—as Polimante calls it—an "empathetic resonance" in his poetry. In discussing the complex connections between speech, body, medium, and social relations in the works of these three poets, Polimante develops a critical framework for the sound-driven exploration of voice and the poetic practice of voicing. **Christof Decker** makes a case for sound as an integral part to film studies in "Practices of Unmixing: Film Aesthetics, Sound and the New Hollywood Cinema." After delineating the privileged position of vision over sound in film history, Decker discusses the practice of sound mixing in Robert Altman's 1975 movie *Nashville*. By focusing on the creative manipulation of noise, background sound, voice, and other sonic elements, he highlights their significance for film aesthetics in a research area primarily concerned with music when engaging with the auditory dimensions of the medium. According to Decker, Altman's mixing, or rather, unmixing of sound in *Nashville* is an experiment in both live sound recording and postproduction that defines the oppositional, countercultural stance of the film.

Part II, "Soundtracks of Collective Memory," retraces the cultural resonances of collective trauma and shared histories of forced displacement, deprivation of rights, and colonial violence. With a focus on the collective memories of Indigenous peoples and postcolonial and diasporic communities across the Black Atlantic,[22] this part explores the acoustics of remembering the past, shaping the present through social transformation, and projecting possible futures. In this context, the term "soundtracks" refers as much to artistic, musical, and literary sonifications of collective memories and the sonic resistance to hegemonial silencing and erasure in cultural practices as to the auditory tracking of cultural reclamation, transnational relations, and interethnic alliances in postcolonial geographies. Such tracking of the roles and functions of sound in cultural engagements with collective memory is inherently multimodal as the following chapters show. The selected case studies from poetry, art, and music consistently take into account the interconnections between sound, vision, and text as they stress the significance of acoustics for the remembrance of shared histories that continue to affect present-day social relations.

In "Voice and Wake: Susan Howe, M. NourbeSe Philip, and the Ecology of Echology," **Julius Greve** explores Philip's and Howe's experimental poetry in a comparative reading of their shared poetic concern with Native American and African diasporic collective memories and historical archives. He delineates the role of poetic form, sound, and voice in the context of histories of forced displacement and genocide that Howe and Philip negotiate and that continue to inform social issues in North America. In a poetological analysis of their works, Greve demonstrates how both poets use acoustic concepts to link the social inscription of colonial violence into the ecologies of land and sea—particularly the Peskeompscut or Turner Falls massacre of 1676 in Howe's "Articulation of Sound Forms in Time" and the 1781 massacre of captured slaves by the crew of the British slave ship *Zong* during the Middle Passage in Philip's poem of the same name—to the temporal persistence of past collective traumas that live on as echoes in the present. **Sabine Kim** writes about the repatriation of First Nations acoustic cultural heritage in "Reframing Indigenous Sonic Archives: Jeremy Dutcher and the Cultural Politics of Refusal." Her chapter offers an analysis of the sociohistorical dynamics behind contemporary artistic efforts to reclaim Indigenous oral culture, song, and ceremonial dance from museum archives. Kim describes the colonial logic behind nineteenth- and twentieth-century anthropological sound collection projects in Canada and their complicity in the cultural suppression of First Nations and the hegemonial erasure of an Indigenous living present. She discusses the musical oeuvre of composer Jeremy Dutcher, who weaves Wolastoqey songs from the archives into his own compositions on his 2018 album *Wolastoqiyik Lintuwakonawa* and uses the audiovisual medium of music videos to embed his songs in the larger social context of the struggle for Indigenous self-determination by adding sound recordings and dance performances that underscore a present-day practice of Indigenous culture. **Tsitsi Jaji** tracks transnational collective connections in contemporary artistic practices in "Unsettled Scores: Listening to Black Oklahoma on the American 'Frontier' " and establishes listening as a means to highlight "potential solidarity" between Indigenous, Black diasporic, and postcolonial African communities. She observes a shared interest in resisting hegemonial politics of social division promoted by racialization and segregation in the works of sculptor Ousmane Houchard Sow and composer and Jazz musician Jacques Coursil who both conceive of the Indian Removal and the Middle Passage as connecting experiences of forced displacement and colonial violence. Jaji understands Sow's 2009 installation *The Battle of Little Bighorn* and Coursil's 2011 album *Trails of Tears* as multimodal works of art that use sound, vision, and text to create spaces of collective listening that invite their audiences to explore potential Indigenous-Black alignments in social justice issues.

"Social Acoustics and Politics of Sound," Part III of the edited collection, pursues intersectional and cultural politics-oriented perspectives on sound in fiction, film, and TV series.[23] Chapters in this section are concerned with the acoustics of racialization, class, and gender and the respective areas of intersection and connection that become audible when critical attention is paid to the politics of sound. The following contributions have a shared interest in showing how fictional storyworlds in the novel and on screen make sense of the interlaced complexity of social conditions and ongoing renegotiations through sound and sonic representation. Together, they demonstrate that the frictions resulting from the combination of acoustic imagination with image and text create multidimensional narratives apt to capture the challenges and potentials of society in the making through the senses.

Gendering and racialization in state communication technology is the dual focus in **Allison Whitney**'s "The Operator and the Final Girl: Gender, Genre, and Black Sonic Labor in *The Call*," Brad Anderson's film from 2013. Whitney contrasts the decades-long Hollywood practice of casting Black actresses as supporting characters in helper roles with the envisioning of a "post-racial collective" that she observes between the two genre-based character types in *The Call*, the horror film's "Final Girl" and the emergency "Operator" figure common to crime narratives. Sound, or rather the characters' collaborative approach to sound, listening, and sensory information, is instrumental in reimagining both roles as they establish an alliance over the telephone. According to Whitney, the Black protagonist's acoustic expertise and her vocal insertion into scenes of violence transcends the disembodiment of her government-assigned role, and challenges the power dynamics of victimization and external determination in the film. **Florian Groß** studies the representational politics of indie music culture in "Bohemian Like You: The Construction of Cool Sound Collectives in Serial Television." He argues that *The O.C.* (Fox, 2003–7), *Californication* (Showtime, 2007–14), and *Mad Men* (AMC, 2007–15) straddle the line between mainstream consumerism and subcultural nonconformism with their soundtracks. His analysis of the complex relationship between the series and alternative rock through the lenses of consumer culture and class elucidates the intricate workings of individual and aesthetic distinction in twenty-first-century capitalism. In "Sonic Sites of Subversion: Listening and the Politics of Place in Karen Tei Yamashita's *Tropic of Orange*," **Nathalie Aghoro** discusses how literary sonification and listening constructs a fictional version of Los Angeles as a city that is defined not only by landmarks or street corners, but also by its polyphonic soundscapes. She argues that the novel captures the everyday dynamics of the city by establishing a tangible, auditory imaginary that reflects its multiethnic formations, class-ridden alignments, and global inflows as much as it challenges and subverts the strict social demarcations that spatiovisual topographies may suggest. The case study understands Yamashita's literary engagement with the soundscapes of Los Angeles as a subversive sonic mapping that unfolds its constitutive power during processes of social change.

The final chapter closes the combined inquiries into *The Acoustics of the Social on Page and Screen*. In "From 'Dead Spots' to 'Hot Spots:' Ann Petry's 'On Saturday the Siren Sounds at Noon,'" **Jennifer Lynn Stoever** draws our attention to the racialized and gendered politics of sound in states of emergency. By establishing connections between the sounding of sirens in Ann Petry's New York City during the Second World War and the Covid-19 pandemic, Stoever maps out how alarm sounds divide the city in protected and neglected areas. She demonstrates that Petry's short story identifies the aural imaginary of sirens as a key site for amplifying racism and other intersecting systemic inequities and revealing their devastating effects on marginalized citizens. According to Stoever, echoes of the hegemonial distribution of political protection in previous times of crisis can still be heard in our contemporary moment.

## Notes

1. This understanding of the term "soundscape" is, and has for a long time been, a departure from Schafer's prescriptive definition of the term, who frames soundscapes as forms of sonic cultural heritage endangered by change and strives to restore and protect them from negative influences like noise pollution. Such a normative approach raises questions of hegemonial and institutional power and requires a critical assessment of who decides which sounds are noise and which ones are worth preserving. C.f. Raymond Murray Schafer, *The Soundscape: Our Sonic Environment and the Soundscape* (Rochester: Destiny Books, 1994).
2. One example of the ideological bias of listening is the "sonic color line," a sonic demarcation along racialized lines that Jennifer Lynn Stoever uses to investigate the acoustic dimensions of "race" beyond the visual with reference to W. E. B Du Bois's concept of the color line. She conceptualizes the listening ear as "the ideological filter shaped in relation to the sonic color line. The listening ear represents a historical aggregate of normative American listening practices and gives a name to listening's epistemological function as a modality of racial discernment" (Stoever, *The Sonic Color Line: Race and the Cultural Politics of Listening* [New York: New York University Press, 2016], 13).

3   Michael Bull and Les Back, "Introduction: Into Sound," in *The Auditory Culture Reader*, ed. Michael Bull and Les Back (Oxford and New York: Berg, 2006), 4.
4   Brandon LaBelle, *Acoustic Territories: Sound Culture and Everyday Life* (New York and London: Continuum, 2010), xxiii.
5   Kara Keeling and Josh Kun, "Introduction: Listening to American Studies," in *Sound Clash: Listening to American Studies*, ed. Kara Keeling and Josh Kun (Baltimore, MD: Johns Hopkins University Press, 2012), 446.
6   Benedict Anderson, *Imagined Communities: Reflections on the Origin and Spread of Nationalism* (London: Verso, 2006), 6. For Anderson the nation is "imagined because the members of even the smallest nation will never know most of their fellow-members, meet them, or even hear of them, yet in the minds of each lives the image of their communion" (ibid.).
7   Cf. Charles Taylor, *Modern Social Imaginaries* (Durham, NC: Duke University Press, 2003).
8   Ibid., 2.
9   Suzi Adams, Paul Blokker, Natalie J. Doyle, John W. M. Krummel, and Jeremy C. A. Smith, "Social Imaginaries in Debate," *Social Imaginaries* 1, no. 1 (2015): 17.
10  Thinking about sound in terms of the work it does is inspired by the conceptual outline of the 2017 conference "Literature, Culture, and the Work of the Humanities" organized by Erica Fretwell (University at Albany, SUNY) and Todd Carmody (Bates College) at the Freiburg Institute for Advanced Studies.
11  Brandon LaBelle, *Sonic Agency: Sound and Emergent Forms of Resistance* (London: Goldsmith Press, 2018), 2.
12  Robin James, *The Sonic Episteme: Acoustic Resonance, Neoliberalism, and Biopolitics* (Durham, NC: Duke University Press, 2019), 3.
13  Ibid.
14  Ibid., 53.
15  Ibid. Social media and smartphone sound cues come to mind as examples for the sonic episteme, since they seek to determine and monitor the user's actions through the unconscious.
16  Ibid., 5–6.
17  Ibid., 6.
18  Alexander G. Weheliye, *Phonographies: Grooves in Sonic Afro-Modernity* (Durham, NC: Duke University Press, 2005), 16.
19  Salomé Voegelin considers listening to be "an activity, an interactivity, that produces, invents and demands of the listener a complicity and commitment" (Voegelin, *Listening to Noise and Silence: Towards a Philosophy of Sound Art* [New York and London: Continuum, 2010], xv). Her definition of listening as a simultaneously personal and reciprocal activity underscores the relational agency that resides in paying attention to sound and illustrates the difference between the practice of listening and the sense of hearing. The commitment that listening requires is intentional and can therefore be granted or withdrawn by tuning out or ignoring social participants on an individual and a collective level. Along these lines, the distribution of listening attention in social contexts raises ethical questions. For Les Back, "[t]o turn a deaf ear is an offence not only to the ignored person but also to thinking, justice and ethics" and he consequently considers the role of the "listener – as the society's ear – [to] establish … an ethical link to those who are not heard or

who are ignored" (Back, "The Listeners: Les Back on the Ordinary Virtues of Paying Attention," *New Humanist Magazine* 125, no. 4 [2010]: n.p.).

20  For sound theories in this context that move beyond established media boundaries and use sound as a critical method that challenges disciplinary demarcations, see: Tsitsi Jaji, *Africa in Stereo: Modernism, Music, and Pan-African Solidarity* (Oxford: Oxford University Press, 2014); Stoever, *The Sonic Color Line*; Weheliye, *Phonographies*; and Carter Mathes, *Imagine the Sound: Experimental African American Literature after Civil Rights* (Minneapolis: Minnesota University Press, 2015). Notably, Mathes frames specific sonic phenomena and characteristics capable of causing friction by establishing a "[r]esistant aurality" in literature as well as in literary criticism, that is, "sonic effects and states such as dissonance, vibration, and resonance . . . [that] bridge the aural, the literary, and the political, complicating ideas and representations of time, memory, and narrative perspective within formal literary studies" (Mathes, *Imagine the Sound*, 1, 6).

21  Nicole Brittingham Furlonge, *Race Sounds: The Art of Listening in African American Literature* (Iowa City: University of Iowa Press, 2018).

22  When Paul Gilroy conceptualizes the transnational, intercontinental connections and cultural networks of Black cultures as the "Black Atlantic," the sounds of Black music play a substantial role in delineating these postcolonial cultural formations as they transgress national and territorial boundaries. C.f. Paul Gilroy, *The Black Atlantic: Modernity and Double Consciousness* (Cambridge, MA: Harvard University Press, 1993).

23  On debates about the critical scope of intersectionality, see: Jennifer C. Nash, "Re-thinking Intersectionality," *Feminist Review*, no. 89 (2008): 1–15; Sumi Cho, Kimberlé Williams Crenshaw, and Leslie McCall, "Toward a Field of Intersectionality Studies: Theory, Applications, and Praxis," *Signs* 38, no. 4 (2013): 785–810.

# Works Cited

Adams, Suzi, Paul Blokker, Natalie J. Doyle, John W. M. Krummel, and Jeremy C. A. Smith. "Social Imaginaries in Debate." *Social Imaginaries* 1, no. 1 (2015): 15–52.

Anderson, Benedict. *Imagined Communities: Reflections on the Origin and Spread of Nationalism*. London: Verso, 2006.

Back, Les. "The Listeners: Les Back on the Ordinary Virtues of Paying Attention." *New Humanist Magazine* 125, no. 4 (2010). https://newhumanist.org.uk/2346/the-listeners.

Bull, Michael, and Les Back. "Introduction: Into Sound." In *The Auditory Culture Reader*, edited by Michael Bull and Les Back, 1–23. Oxford and New York: Berg, 2006.

Cho, Sumi, Kimberlé Williams Crenshaw, and Leslie McCall. "Toward a Field of Intersectionality Studies: Theory, Applications, and Praxis." *Signs* 38, no. 4 (2013): 785–810.

Furlonge, Nicole Brittingham. *Race Sounds: The Art of Listening in African American Literature*. Iowa City: University of Iowa Press, 2018.

Gilroy, Paul. *The Black Atlantic: Modernity and Double Consciousness*. Cambridge, MA: Harvard University Press, 1993.

Jaji, Tsitsi. *Africa in Stereo: Modernism, Music, and Pan-African Solidarity*. Oxford: Oxford University Press, 2014.

James, Robin. *The Sonic Episteme: Acoustic Resonance, Neoliberalism, and Biopolitics*. Durham, NC: Duke University Press, 2019.
Keeling, Kara, and Josh Kun, eds. *Sound Clash: Listening to American Studies*. Baltimore, MD: Johns Hopkins University Press, 2012.
LaBelle, Brandon. *Acoustic Territories: Sound Culture and Everyday Life*. New York and London: Continuum, 2010.
LaBelle, Brandon. *Sonic Agency: Sound and Emergent Forms of Resistance*. London: Goldsmith Press, 2018.
Mathes, Carter. *Imagine the Sound: Experimental African American Literature after Civil Rights*. Minneapolis: Minnesota University Press, 2015.
Nash, Jennifer C. "Re-thinking Intersectionality." *Feminist Review*, no. 89 (2008): 1–15.
Schafer, Raymond Murray. *The Soundscape: Our Sonic Environment and the Soundscape*. Rochester: Destiny Books, 1994.
Stoever, Jennifer Lynn. *The Sonic Color Line: Race and the Cultural Politics of Listening*. New York: New York University Press, 2016.
Taylor, Charles. *Modern Social Imaginaries*. Durham, NC: Duke University Press, 2003.
Voegelin, Salomé. *Listening to Noise and Silence: Towards a Philosophy of Sound Art*. New York and London: Continuum, 2010.
Weheliye, Alexander G. *Phonographies: Grooves in Sonic Afro-Modernity*. Durham, NC: Duke University Press, 2005.

Part I

# Sound Practice across Media

# 1

# Listening in Print[1]

## Nicole Brittingham Furlonge

When accepting the National Book Foundation medal for distinguished contribution to American letters on November 6, 1996, Toni Morrison reflected on peace. "There is a certain kind of peace that is not merely the absence of war," Morrison begins. "The peace I am thinking of," she continues, "is the dance of an open mind when it engages another equally open one – an activity that occurs most naturally, most often in the reading/writing world we live in."[2] She cautions that the peace emerging from such a dance needs to be intentionally secured and "warrants vigilance"; it is not one to take for granted.[3] For Morrison, the fast pace of our busy culture endangers such peace. So, too, does "the physical danger to writing suffered by persons . . . who live in countries where the practice of modern art is illegal and subject to official vigilantism and murder."[4] Her metaphor of the dancing mind envisions such intentional work as vital collaborative co-creation across difference: "Its real life is about creating and producing and distributing knowledge; about making it possible for the entitled as well as the dispossessed to experience one's own mind dancing with another's."[5] Morrison's metaphor is one through which knowledge becomes accessible to all. Knowledge, then, is not a thing acquired. Instead, it is about intentional engagement, process, action, meaning making and meaning sharing in relation to another. For Morrison, how we engage in the reading and writing life is an essential, relational dance.

It is not surprising that Morrison offers a musically infused metaphor—a dancing mind—to describe the dynamic and vital relationship between readers, writers, and the pages of texts. Dubbing Morrison "a rebel sister theorist of music," Daphne Brooks reminds us of Morrison's musicianship; along with fiction and essays, she authored lyrics and operas that demonstrated her understanding and valuing of "music as an insurgent expression of black interiority" and "one of the most potent forms we have to dissect America's racial complexities and to affirm the prodigious expanse of black humanity."[6] Morrison was deeply committed to "restoring the articularity of sound" to print and to the literary form of the novel in particular.[7] Her aesthetic was one that worked to create the kinds of novels that would possess "a non-book quality, so that they would have a sound . . . like somebody was telling you a story."[8] Rooting her writing in oral storytelling traditions as well as music, Morrison was invested in creating speakerly literature; she "wanted the sound to be something [she] felt was spoken and more oral and less print."[9] On the printed page, music was an essential part of this sonic restoration project. So, too, was the possibility Morrison heard in language printed on the storytelling page.

If we extend the musical and performative metaphor of the dancing mind further, we become attuned to the reader's need to engage texts with a *listening* mind as well. Or, as I explore in my book *Race Sounds: The Art of Listening in African American Literature*, to consider the possibilities of *listening in print*. *Race Sounds* amplifies listening as a multimodal, relational, contextual, fully embodied dynamic continuum of sonic practices that engage actively in meaning making.[10] As such, *Race Sounds* and the listening it demonstrates self-reflexively question how positionality—specifically race, class, and gender positionality—influence the ways in which we are able to make sense of sound in all the ways it is transmitted to and through us. In this project, a reader, writer, or cultural participant adopts the positionality of a listener who listens *as*, *with*, and *within*. Listening functions as an aural form of agency, as an essential practice of citizenship, as aural empathy, as an ethics of community building, as social action, as cultural revision strategy, and as a practice of historical thinking and witnessing. The specific discourses of African American literary studies, Black Feminist literary and cultural studies, and Sound Studies intersect in *Race Sounds* in order to, as I put it, "enliven how we read, write, and critique texts but also to inform how we might be more effective audiences for each other and against injustice in our midst."[11]

In this chapter, I listen back to *Race Sounds*, tuning in again to the multifaceted emergent practice of listening that I examined in that project, particularly as it plays on the literary page. Here, I explore further the aural practices that *listening in print* engages through sampling (read/hear: like sampling in hip-hop) a few ways in which writers create space for readers to listen. These samples do not weave together to suggest one correct way to listen. Instead, they are offered as sonic sites through which diverse and productive outcomes of a dynamic range of listening practices and their possibilities can emerge.

## Sample 1: Listening in Print as a Discrepant Practice

If, as Nathaniel Mackey asserts, "The page and the ear coexist. Not only do they coexist, they can contribute to one another," then how does one read as a listener?[12] In *Race Sounds*, I coined the phrase *listening in print* in order to describe the listening practices that work to allow "the page" and "the ear" to coexist and contribute to each other's expressivity and possibility. Key to the practices of *listening in print* is the work of unmuting print in order to listen—or tuning to the other side of *printed* language. Philosopher Gemma Corradi Fiumara refers to listening as "the other side of language," the side that we "tend to ignore" because Western thought is predominantly focused on speech.[13] Fiumara's project argues for the need to understand logic—the Greek *logos*—more fully, for "there could be no saying without hearing, no speaking which is not also an integral part of listening, no speech which is not somehow received."[14] This fuller understanding of *logos* is crucial, for instance, to the health of a society. As Fiumara suggests, "if we are apprentices of listening rather than masters of discourse, we might perhaps promote a different sort of coexistence among humans: not so much in the form of a utopian ideal but rather as an incipient philosophical solidarity capable of envisaging the common destiny of the species."[15] We will return to this idea in the conclusion of this chapter.

While other literary traditions engage in sonic practices, I chose to focus in my book on African American literary and cultural texts particularly because, as Robert Stepto asserts, Black texts persuade readers to act as "storylisteners," that is, "to seek the kind of communal relationship found, for example, between preachers and congregations, musicians and audiences in certain performance venues, and between storytellers and storylisteners."[16] Here, Stepto situates literary texts alongside particular Black sonic cultural practices in dynamic sites (specifically the church and the theater) that amplify the power of communal meaning-making and knowledge sharing. By extension, rather than a solitary practice, reading, then, claims a rootedness in relational creative practices. In this paradigm, "readers become hearers, with all that that implies in terms of how one may sustain through reading the responsibilities of listenership as they are defined in purely performative contexts."[17] In amplifying the importance of the story*listener*, Stepto—in resonance with Toni Morrison and other Black writers discussed in *Race Sounds*—positions the printed text as full of sonic possibility.

Bruce R. Smith, however, reminds us that even when we consider print on its own, we need to remember that literacy has not always been imagined or practiced as a mute, inflexible, "monolithic entity."[18] In his study of print culture in the early modern world, Smith explains that "where twenty-first-century students are likely to see only marks imprinted on paper – or, ignoring the imprintedness entirely, the concepts that those marks encode," early modern readers "would . . . have heard traces of sound" in "woodcut illustrations, in handwriting, and in print."[19] Even when we read silently to ourselves, Elaine Scarry explains, "the spoken words are acoustically imaged rather than actually heard."[20] Print, then, calls on the reader to not just see, but also to tune in as a multisensory being to the sounds embedded in print materials—and particularly to literature that insists on the intimate and explicit, rather than separate or implicit, relationship between print and sound.

By extension, listening is not a fixed, monolithic way of sensing and making sense. Instead, practicing reading in this aural manner—a *listening in print*—means remaining mindful of what Mackey refers to as discrepancies between print and the sonic life of words. As Mackey explains, discrepant engagement operates "in the interest of opening presumably closed orders of identity and signification, accent fissure, fracture, incongruity, the rickety, imperfect fit between word and world. Such practices highlight—indeed inhabit—discrepancy, engage rather than seek to ignore it."[21] It is useful here, too, to note the Latin *discrepāre*, to crack or creak. Rooted in the word "discrepancy," then, is a relationship between the word/print and the sound of print.

In his epistolary novel *Djbot Baghostus's Run*, Mackey's narrator, N., models discrepant engagement as a process of sense making. N. writes a letter in which he conflates recorded music with his "inventory of traces," a record of his own becoming.[22] He begins compiling this inventory by listening to Miles Davis's *Seven Steps to Heaven*, one of the older albums in his collection. As he listens, N. hears not only Davis' music, "one of the cuts which made me"; he also attends to the "places where the needle skips" on the vinyl, interpreting them not as "noisy reminders of the wear of time," but "as rickety, quixotic rungs on a discontinuous ladder – quixotic leaps or ellipses."[23] *Seven Steps to Heaven* functions not only as a vinyl archive, a recorded holding place for Davis' musical creations which can be accessed repeatedly over time and in various spaces; this vinyl recording also contains, in a sense, N.'s personal evolution and positionality over time. This personal evolution in turn is informed by and perceived through how N. listens to the music. His sense of becoming is mapped through listening and involves the deciphering of the recorded music stored on vinyl, the traces of other moments spent listening to the record, the growing awareness, knowledge, questions, and experiences the listener brings to each opportunity to listen, and the uncertainty inherent in such cultural engagement. As a listener, N. demonstrates his ability to oscillate between layers of sound, discerning the familiar, detecting those utterances that are unfamiliar, and developing listening strategies to engage these multilevel, stacked utterances.

This moment in Mackey's novel also provides an opportunity for us to consider how we come to listen to Miles Davis—or any musician and their music—through a reference in a literary text. For some of my students, reading this passage from *Djbot Baghostus's Run* marks their introduction to the iconic Davis. Mackey's metaphor of the rickety ladder leads them to imagine how *Seven Steps to Heaven* might sound—full of leaps, missteps, and uncertain steps. That, coupled with the sound of the skipping needle on the worn record, provides what they take as clues to decode the novel as a record, too, to be listened to. This album is a noteworthy choice for Mackey to play in this moment in the novel and in light of his discrepant poetics. *Seven Steps to Heaven* is a pivotal album for Davis, one that carries its own inventory of traces in marking just one of Davis' innovative jazz transitions into bebop. The album was also recorded and then re-recorded with new personnel, marking its production with a series of traces of layered expressivity.[24]

The notion of an "inventory of traces" points to identity and text as a compilation of sonic layers, recalling for me J. Martin Daughtry's metaphor of an acoustic palimpsest. In the Middle Ages, palimpsests were manuscripts written on papyrus or vellum that were often washed off and written over. In the process, faint traces of the previous writing remained. The palimpsest is thus the result of successive acts of partial erasure and inscription, acts that turn it into a "multilayered record,"[25] a trace of multiple histories and multiple authors. Daughtry wonders if the palimpsest—a figure he extends beyond the Middle Ages to literary texts and the oft painted over walls of buildings in a city—can be "wired for sound" in order to explore listening as accessing layers of sound, including those otherwise missed or ignored.[26] Mackey's "inventory of traces," rickety ladder, and discrepant engagement, then, are all figures of an acoustic palimpsest. What I appreciate additionally about Mackey is that I hear in these figures the role one's identity and positionalities play as one listens. And that positionality is not fixed, but is instead made up of traces, tracks, layers, and complex histories.

There is a layering of different, discrepant sounds in this passage, positioning the listener-reader to be mindful that the work of the page and the ear, then, is not one kind of listening, nor does it produce one fixed or unified meaning. Instead, listening in print is a noisy, relational, layered generative process that takes place between the listener and the listened-to. Listening in print also tugs on the certain and uncertain positionalities we individually draw from that inform our listening practices. As Mackey's rickety ladder suggests, this listening emerges from an awareness of and relationships between the layers to which we must attend as listeners. Yet the goal is not to attend to all steps equally, nor to resolve the noise. Instead, this practice of listening keeps us mindful of the layers of sounds we are called on to sense as well as the multiple meanings that emerge through these layers.

## Sample 2: The Word and the Sound

*Jazz* remains my favorite of Morrison's novels (if I am pressed to choose a favorite) because it questions its own story's validity, engaging variably in a performance of certainty and doubt. In this self-reflective pose, the novel reminds me of Zadie Smith's essay "Fascinated to Presume: In Defense of Fiction": "Fiction suspected that there is far more to people than what they choose to make manifest . . . Fiction— at least the kind that was any good—was full of doubt, self-doubt above all. It had grave doubts about the nature of the self."[27] From *Jazz*'s first utterance—"Sth, I know that woman"[28]—we become the audience to the narrator's certainty, to her claim of knowledge and, therefore, power. The utterance, "Sth," sonically punctuates what completes the opening sentence: "I know that woman." "Sth" is the sound of sucking teeth, a vernacular punctuation of certainty and expression of an "of course, why would you even question me" attitude. Because of the narrator's knowledge, we should pay attention. Through her certainty, we should feel at ease as readers, listening in print with confidence to her narrative that will emerge from her positionality as one who knows. As a sound, "Sth" invites us as readers to listen and discern sound in print—the layers of print's expressivity—from the beginning.

The City (understood to be New York City) serves as the stage for her certainty. The City in 1926 becomes its own actor in this novel: "Nobody says it's pretty here; nobody says it's easy either. What it is is decisive, and if you pay attention to the street plans, all laid out, the City can't hurt you."[29] Like this City that will do no harm if you heed its design, heeding the voice of the narrator and the form of her narration presumably will yield us as readers similar protection. Beginning with this sonic utterance elicits an interactive, supportive relationship between the page and the ear, book and reader.

"Sth" also marks the space between "the name of the sound / and the sound of the name," two of the lines included in *Jazz*'s epigraph. The epigraph is an excerpt from the poem, "Thunder, Perfect Mind," from the ancient Egyptian codices *The Nag Hammadi*. These poetic lines amplify the slippage between the presumed silence of print and the sound of language. The passage juxtaposes orality ("the sound of the name") and inscription ("the name of the sound"), signifying a space into which

Morrison invites her readers to listen in print. There is an insistence not only that the printed word could sound but also that the responsibility on the part of the reader is to learn the equally myriad, improvisational, strategic ways one could practice listening and, thereby, engage in situated meaning making in the text, on the page.

Morrison's sonic restoration project is resonant in *Jazz*. The book is full of vibrating spaces, migrating men and women dancing into the City on trembling trains, audible (both heard and silenced) touches that slice, murder, and remind, and the horror of diminishing and diminished feeling. Scholars have highlighted aptly the specifically musical elements of *Jazz*, with most commentary focusing on improvisational and repetitive aspects of the text.[30] Yet what strikes me about this novel is its focus on representing jazz—the music, style, and the novel—as a creative act. By extension reading becomes a creative practice. The form and approach to jazz in this vein informs how we move forward to listen as a continuum of practices and processes. I am taken with the ways in which Morrison's *Jazz* constructs a text-based, text-bound narrative space inclusive of sonic listening practices. Jazz music permeates the air in *Jazz*. It cannot be contained or rigidly represented. It is slippery and permeable. You can imagine the sounds resonating from the space of the page, compelling your reading ears to consider the nature of sound to vibrate, resonate, and travel ubiquitously within, around, and through a space—even the space of the page:

> [Y]ou could hear it everywhere. Even if you lived, as Alice Manfred and the Miller sisters did, on Clifton Place, with a leafy sixty-foot tree every hundred feet, a quiet street with no fewer than five motor cars parked at the curb, you could still hear it, and there was no mistaking what it did to the children under their care – cocking their heads and swaying ridiculous, unformed hips.[31]

While the residents of Clifton Place may desire to dampen what, to their ears, are unwelcome, salacious jazz sounds, the music is all over the page. Children dance in the space of the page. It even, figuratively speaking, changes the weather: "Up there, in that part of the City—which is the part they came for—the right tune whistled in a doorway or lifting up from the circles and grooves of a record can change the weather. From freezing to hot to cool."[32]

Music percussion as protest sounds in this novel's Jazz Age and Harlem Renaissance soundscapes as well, complicating the sounds we associate more readily with that era. For instance: "Now, down Fifth Avenue from curb to curb, came a tide of cold black faces, speechless and unblinking because what they meant to say but did not trust themselves to say the drums said for them, and what they had seen with their own eyes and through the eyes of others the drums described to a T."[33] Morrison stunningly constructs a "speechless" space into which we are pulled to listen. The percussion of a steady, resolved, expression-full drum beat breaks through the silence we hear on the page. We listen in the midst of this acoustics of the unspoken to the sound of the drums recorded in print on the page, the trauma of racial violence in the percussive instrument play. Here, the silence of the protesters punctuated with the steady utterance of drums asserts that we should not assume print as mute. Print is, instead, a sonic figure, in need of listeners to attend to its various frequencies.[34]

We may believe in the confident authority of our narrator from the novel's beginning, especially given all the reasons she provides to undergird her knowing: she shares that she and others in Harlem know who Violet is because, "[b]ut, like me, they knew who she was, who she had to be, because they knew that her husband, Joe Trace, was the one who shot the girl."[35] How they know her, what they know of her, that they "know that woman," is based on what makes sense to them. For all the insistence of certainty here (also a certainty I presume in "Sth" because of how I hear and decode this sound), there are moments when the narrator expresses doubts about her knowing. In the final pages of the novel, we listen as the narrator reflects on her work as storyteller: "I break lives to prove I can mend them back again. And although the pain is theirs, I share it, don't I? Of course. Of course. I wouldn't have it any other way. But it is another way. I am uneasy now. Feeling a bit false. What, I wonder, would I be without a few brilliant spots of blood to ponder? Without aching words that set, then miss, the mark?"[36] In this writerly moment, the narrator questions her very role and the story it produced. Doing so through the mode of inquiry and still in conversation with the reader as listener, the narrative models what it might sound like for a reader to pause and wonder not only how they journeyed to this moment as a reader, but also how they were listening and what their *listening in print* produced.

In the spirit of pausing, wondering, and reflecting, let's return for a moment to the beginning of the novel. If we decouple "Sth" for a moment from critics and even Morrison's prevailing explanation of this utterance as sucking teeth, we might also detect a different meaning possibility. For instance, I listen again to "Sth" and imagine the sound of a typewriter's typebar striking paper and making a mark. Aurally imagined as such, "Sth" also makes audible the process and very act of writing or producing the book—the work of creating, producing, and disseminating knowledge that Morrison reflects on in "The Dancing Mind."

Back to the closing pages of the novel, our narrator continues: "I ought to get out of this place. Avoid the window; leave the hole I cut through the door to get in lives instead of having one of my own. It was loving the City that distracted me and gave me ideas. Made me think I could speak its loud voice and make that sound sound human. I missed the people altogether . . . Now I know."[37] I listen to this statement of knowing quite differently from the one with which we begin the novel. *Jazz*'s narrator progressively exposes her desire for listening, until she makes her way to a firm imperative by the end of the novel:

> But I can't say that aloud; I can't tell anyone that I have been waiting for this all my life and that being chosen to wait is the reason I can. If I were able I'd say it. Say make me, remake me. You are free to do it and I am free to let you because look, look. Look where your hands are. Now.[38]

We conclude with an imagined invitation for readerly-writerly collaboration, with a passage that calls out a desire for the book to be able to speak and the reader to be able to listen. Here, touch and sound commingle in the possibility that, if the book could speak, if the reader could hear the narrator so that the transmission of the message could occur, then the reader could engage differently. Rather than fixed certainty or narrative failure, Morrison posits here in this intimate moment, on the last page of the novel, other possibilities: Perhaps we can awaken our full capacity for listening, for sensation, to the page, the story, and to the people and spaces within it. Perhaps we might listen more dynamically for the possibility of relationship between us as listeners and the things with which we interact and attempt to sense. Perhaps we can understand the book as a dynamic set of printed processes rather than a text fixed in print. As Daughtry suggests about the consideration of music through the metaphor of the acoustic palimpsest, "[t]he effect of this move is to blur the line between the musical object and the sonorous world; to allow the cacophony of the world to rush into the study of music; and to place the politics of navigating through this complex and noisy world at the center of discussions of listening."[39] I suggest that, through the practice of listening in print, such cacophony finds its way into the world of the book as well. As such, encountering such multilayered opportunities to listen in print helps us understand the page as practice space, one in which we can grow our capacity for the essential ethical and multilayered listening our world needs.

## Sample 3: Trouble in Mind and Inner Listening

Morrison does not dance alone in her desire to create such books that invite—even provoke—us to listen as we read. Like Morrison—and Stepto—Gayl Jones crafts narratives that are grounded in a sense of story*tellers* and story*hearers*. Jones describes herself as storyteller as opposed to fiction writer, explaining that: "When I say 'fiction,' it evokes a lot of different kinds of abstractions, but when I say 'storyteller,' it always has its human connections . . . There is always that kind of relationship between a storyteller and a hearer – the seeing of each other. The hearer has to see/hear the storyteller, but the storyteller has to see/hear the hearer."[40] Jones' treatment of storytelling as relational exchange extends to her fiction. In her first novel, *Corregidora*, Jones presents Ursa Corregidora, a blues vocalist who negotiates a continuum of historical, familial, and personal creative trauma that is shared between characters through storytelling. While the novel is set in Kentucky in the mid-1940s to the late 1960s, the historical afterlife of slavery,[41] particularly the experiences of enslaved women in nineteenth-century Brazil, is palpable still in the lives of the Corregidora women. Ursa is the fourth generation of Corregidora women and grows up with the matriarchal edict to bear witness to slavery's atrocities by giving birth to a (female) child and passing on orally the evidentiary narrative to her offspring. The traumatically repeated concern of the Corregidora women is insistent in part because, despite the abolition of slavery in Brazil, the control over future generations was contestable, particularly because the state burned all evidence of the institution's existence. There is no account for lived life, then, only for that which is spoken and listened to. The novel speaks simultaneously to historical horror and personal loss primarily through her foremothers with Ursa serving as their audience.

We first *listen in print* as Ursa sits in the intimate space of her room, vocalizing for the first time since her hysterectomy. She hums a portion of the song "Trouble in Mind"[42]—"the part about taking my rocking chair down by the river and rocking my blues away."[43] She then reflects, "What she [Cat] said about the voice being better because it tells what you've been through. Consequences. It seems as if you're not singing the past, you're humming it. Consequences of what? Shit, we're all consequences of something. Stained with another's past as well as our own. Their past in my blood."[44] Ursa's voice is both strained and stained, already part of the evolving historical and familial narrative. Rather than Ursa's complete erasure from the record, then, Ursa's miscarriage and subsequent hysterectomy serve as impetus for change in her blues art, her relationship to her maternal forebears and their narrative, and her position as witness—a listener to the past and aural agent in the present. Further, in this moment of humming "Trouble in Mind," Ursa calls attention to a distinction between "singing"—the voice of the vocalist traveling beyond the body to her audience—and "humming"—the voice heard by an audience, but reverberating in the body of the vocalist. Here, the voice of the vocalist is inscribed on the page of the novel as well as marked with the multilayered historical and current experiences that Ursa bears. As readers, we are audience to not only Ursa's voice in this moment, but also to her own listening. That listening attunes our reading ears to her interior acoustics made public, or published.

This tuning to an interior acoustics made public is important for it allows Ursa to filter her foremothers' process of archiving the trauma of slavery. Ursa's grandmother insists: "*They burned all the documents, Ursa, but they didn't burn what they put in their minds. We got to burn out what they put in our minds, like you burn out a wound. Except we got to keep what we need to bear witness. That scar that's left to bear witness. We got to keep it as visible as our blood.*"[45] According to Gram, for the Corregidora women, the trauma of slavery must be internalized and branded on their "conscious" through perpetual, imperative storytelling. Emancipation is insufficient to remove the suffering, sexual exploitation, and incest suffered in and after slavery and leads the Corregidora women to create another scar, seared into memory and the "conscious" in order to generate intergenerational memory as evidence.

While Ursa's foremothers also point to the significance of the internal record of the mind—the socio-emotional, psycho-social archive—their conceptualization of the interior is quite different from Ursa's emerging interior acoustics made public. For gram and great-gram, as long as the record of that original historical and traumatic memory passed on is intact, evidence remains. What they also assert is that what was put in the slave's mind, their "trouble in mind," what was internalized of the harm done, should be burned out like a wound. The language of healing here is both provocative and invasive. One burns a wound to sterilize it, close it, and help it heal. So, the call is to burn out the harm, the pain, and to keep the scar as record—the left over, what remains. Evidence of historical atrocity, then, becomes physical and material—blood, visible body, *scar* on the mind. But it is also psychological and psychical—a scar *on the mind*, a personal experience internalized as historical record. The destroyed legal record resurfaces here both as spoken and as indelible evidence on the mind.

For the Corregidora women, the perpetuation of this internal evidence also takes the form of offspring, creating conscious beings to retell and maintain the evidence until it comes time to produce it: *"The important thing is making generations. They can burn the papers but they can't burn conscious, Ursa. And that what makes the evidence. And that's what makes the verdict."*[46] The offspring become "conscious" multiplied, making the scar visible and, as interlocutors, amplifying the scar's audibility. With Ursa's inability to bear children, then, comes a break in the family's trail of storytelling evidence: "My great-grandmama told my grandmama the part she lived through that my grandmama didn't live through and my grandmama told my mama what they both lived through and my mama told me what they all lived through and we were supposed to pass it down like that from generation to generation so we'd never forget . . . Yeah, and where's the next generation?"[47] For Ursa, this is a poignant and high-stakes question. Without the next generation, Ursa does not have children to pass the story on to or to bear witness for her.

The crisis of witnessing does not stop there. When discussing her mother, Correy (her name a diminutive of Corregidora and a name her husband, Martin, used to call her),[48] Ursa wonders at the *interior* story Correy never told her: "*Mother still . . . carried their evidence, screaming, fury in her eyes, but she wouldn't give me that, not that one. Not her private memory.*"[49] As Ursa tries to figure out other possibilities for herself in terms of her responsibility to bearing witness, she visits Correy, longing to talk with her: "I could feel the strain and wondered if she could. I'd always loved her and knew she loved me, but still somehow we'd never 'talked' things before, and I wanted to talk things now."[50] This moment of talk between mother and daughter carries with it the figure of the strain that Ursa often mentions in relation to her post-surgery singing voice and a longing for intimate dialogue.

During this visit home, Correy shares parts of her own story with Ursa, particularly her experiences with Ursa's father, Martin. Correy becomes pregnant after her first sexual encounter with Martin; Great-Gram then forces Martin to marry Correy and they live together in the same house with Gram and Great Gram. As Correy describes her life with Martin, her story emerges "in pieces, instead of telling one long thing."[51] While much of her story focuses on her time with Martin, Correy slips between her story and that of Great-Gram and Gram, interweaving the two narratives as if they are one. More striking, according to Ursa, Correy also shifts from *herself* telling her own story and morphs into what seems to Ursa like *Great Gram* telling *her own* story: "Mama kept talking until it wasn't her that was talking, but Great Gram. I stared at her because she wasn't Mama now, she was Great Gram talking."[52] Although Correy is able to share portions of her "own private memory" with her daughter, she is unable to disentangle her memory completely from that of her foremothers. Her storytelling leaves Ursa noting that "it was as if their memory, the memory of all the Corregidora women, was her memory too, as strong with her as her own private memory, or almost as strong."[53] Correy also stops short of sharing the circumstances of her current life with her daughter, leaving Ursa wanting "to ask what about her now, how lonely was *she*. She'd told me about *then*, but what about *now*."[54] The punctuation here—periods instead of question marks—underlines that these are questions left unspoken.

While Ursa is left wondering after her conversation with Correy, we still listen and perceive Correy's decision not to share certain experiences with Ursa as an empowering choice. As Kevin Quashie reminds us, while African American culture is often considered expressive, dramatic, and even defiant, quiet can be heard here as a different kind of expressiveness, particularly one which speaks to Correy's vulnerabilities and fears. Strategic silence, as Quashie explains, "is often aware of an audience, a watcher or a listener whose presence is the reason for the withholding."[55] While Correy strategically withholds information about her current life, she plants the seed for questioning, for the possibility of an inheritance that differs markedly from the evidence-building narrative of her Great-Gram and Gram, a narrative that makes no room for questions. We listen as Ursa shares that once, when Ursa was still a child, Correy questioned her mother's and grandmother's narrative: "*How can it be?*" Ursa reflects, "[Correy] *was the only one who asked that question, though. For the others it was just something that was, something they had, and something they told. But when she talked, it was like she was asking that question for them, and for herself too.*"[56] Posing questions is quite a different listening habit than Ursa was raised to practice. Growing up, hers was a rote listening in which posing questions was not only discouraged, but also punished.[57] Correy's question "How can it be?" signals her openness to listen toward another possibility, specifically toward an altered storytelling and storylistening practice.[58]

After visiting with her mother, Ursa reflects: "*[S]till it was as if my mother's whole body shook with that first birth and memories and she wouldn't give those to me, though she passed the other ones down, the monstrous ones, but she wouldn't give me her own terrible ones . . . How could she bear witness to what she'd never lived, and refuse me what she had lived? . . . What was their life then? Only a life spoken to the sounds of my breathing or a low-playing Victrola . . . What's a life always spoken, and only spoken?*"[59] Ursa, in her quiet questioning, is defiant in her own way. Her question troubles the narrative repetition of the Corregidora women. Understood is the implication that a life only spoken and not lived is not a full life. I would add, however, that Ursa's question also gestures toward a lack of conscious listening to the life repeatedly spoken. That is, the Corregidora women live as if a spoken, witnessing voice is proof enough—all at the expense of a cultivated personal and interior life.

Ursa's questioning and listening here is necessarily discerning and signals that the listening practice she now brings to bear is one that does not accept narratives without reflecting on them. She instead embraces in her listening the need to filter and discern what she hears as she determines the new story she wishes to tell and sing.

## Pause

*How might we be better audiences for each other?* This question encapsulates why I am focused on the continuum of essential habits that are listening. As Susan Bickford asserts, listening is "a central activity of citizenship."[60] In *Talking to Strangers*, Danielle Allen identifies interracial distrust as what congeals at the core of citizenship and democratic problems in the United States. For Allen, "On the same page or in the same city ... citizens of different classes, backgrounds, and experiences are inevitably related to each other in networks of mutual benefaction ... This relationship is citizenship, and a democratic polity, for its own long-term health, requires practices for weighing the relative force of benefactions and for responding to them."[61] While Allen does not speak explicitly about civic listening, her study does point to listening as an ethical practice of social, public, and civic citizenship. The necessary work of listening in conversation between stranger citizens is a way of developing habits that can improve the workings of democracy. From such democratic listening can emerge a curative space, one where rupture is named, aired, and potentially repaired. Such listening might lead us toward realizing intersectional justice.[62]

We rewind, as promised, to Fiumara and the other side of language. Because misunderstanding, according to Fiumara, is "deeply rooted in the exclusion of listening," the work of listening is central to the creation of the conditions necessary for true dialogue and exchange, the conditions necessary for the generation of new knowledge, understanding, and repair.[63] This is a notion of listening *as* justice, a listening that stretches toward a kind of justice. It is in the relationship between writer, listener, and context—in the ethical practice of *listening in print*—that emergent, new worlds and possibilities for new, more humane citizen relationships reside.

## Notes

1. This contributed chapter was adapted from portions of *Race Sounds: The Art of Listening in African American Literature* by Nicole Brittingham Furlonge (Introductions and Chapter 3) © 2018 University of Iowa Press. Used with permission of the University of Iowa Press.
2. Toni Morrison, *The Dancing Mind* (New York: Knopf, 1996), 7.
3. Ibid.
4. Ibid., 14.
5. Ibid., 16.
6. Daphne A. Brooks, "Toni Morrison and the Music of Black Life," *Pitchfork*, August 15, 2019. https://pitchfork.com/thepitch/toni-morrison-and-the-music-of-black-life/.
7. Ibid.
8. Quoted in Cheryl Hall, "Beyond the 'Literary Habit': Oral Tradition and Jazz in Beloved," *MELUS* 19, no. 1 (1994): 89.
9. Ibid.
10. This idea of and amplifying of listening is not dependent on one's ability to physically hear. See Steph Ceraso's discussion of the musical practice of Evelyn Glennie, a hearing-impaired musician who offers listening as a fully embodied, vibrational practice, in *Sounding Composition: Multimodal Pedagogies for Embodied Listening* (Pittsburgh, PA: University of Pittsburgh Press, 2018).
11. Furlonge, *Race Sounds*, 17.
12. Nathaniel Mackey, *Paracritical Hinge: Essays, Talks, Notes, Interviews* (Madison: University of Wisconsin Press, 2005), 313.

13  Gemma Corradi Fiumara, *The Other Side of Language: A Philosophy of Listening*, trans. Charles Lambert (New York and London: Routledge, 1990), 1.
14  Ibid.
15  Ibid., 57.
16  Robert Stepto, *A Home Elsewhere: Reading African American Classics in the Age of Obama* (Cambridge, MA: Harvard University Press, 2010), 151.
17  Ibid., 145.
18  Bruce R. Smith, *The Key of Green: Passion and Perception in Renaissance Culture* (Chicago: University of Chicago Press, 2009), 32.
19  Ibid., 23.
20  Elaine Scarry, *Dreaming by the Book* (Princeton, NJ: Princeton University Press, 2001), 132. See also Andrea Moro's piece on the sound of silent thought when we read to ourselves in relation to how our brain processes our reading out loud: "What Is the Sound of Thought?" *The MIT Press Reader*, September 18, 2016, https://thereader.mitpress.mit.edu/the-sound-of-thought/.
21  Nathaniel Mackey, *Discrepant Engagement: Dissonance, Cross-Culturality, and Experimental Writing* (Cambridge: Cambridge University Press, 1993), 19.
22  Nathaniel Mackey, *Djbot Baghostus's Run* (Los Angeles: Sun and Moon Press, 1993), 150.
23  Ibid.
24  See https://www.jazziz.com/miles-davis-seven-steps-to-heaven/.
25  Oxford English Dictionary.
26  J. Martin Daughtry, "Acoustic Palimpsests and the Politics of Listening," *Music and Politics* 7, no. 1 (2013): 1–34. This figure of the acoustic palimpsest recalls, too, Ralph Ellison's ethical concern with tuning in to those in history who are less perceptible to the societal ear. See my reading of *Invisible Man* in *Race Sounds*.
27  Zadie Smith, "Fascinated to Presume: In Defense of Fiction," *The New York Review*, October 24, 2019, https://www.nybooks.com/articles/2019/10/24/zadie-smith-in-defense-of-fiction/.
28  Toni Morrison, *Jazz* (New York: Plume, 1993), 3.
29  Ibid.,, 8.
30  See, for instance, Roberta Rubenstein, "Singing the Blues/Reclaiming Jazz: Toni Morrison and Cultural Mourning." *Mosaic: An Interdisciplinary Critical Journal* 31, no. 2 (1998): 147–63.
31  Morrison, *Jazz*, 56.
32  Ibid., 51.
33  Ibid., 54.
34  This reading is resonant with Jean Luc Nancy's notion of silence as situated in and still within sound. See *Listening* (New York: Fordham, 2007), 21.
35  Morrison, *Jazz*, 4.
36  Ibid., 219.
37  Ibid., 1.
38  Ibid., 229.
39  Daughtry, "Acoustic Palimpsests," 11.
40  Michael Harper and Robert B. Stepto, *Chant of Saints: A Gathering of Afro-American Literature, Art, and Scholarship* (Urbana: University of Illinois Press, 1979), 357.

41. Saidiya Hartman, *Lose Your Mother: A Journey Along the Atlantic Slave Route* (New York: Farrar, Straus, and Giroux, 2007), 6.
42. "Trouble in Mind" is a blues standard, performed and recorded by many artists. The version I wish to note here is that of Nina Simone (1960): https://www.youtube.com/watch?v=UZmPWvAiNs4.
43. Gayl Jones, *Corregidora* (Boston, MA: Beacon Press, 1975), 45.
44. Ibid.
45. Ibid., 72.
46. Ibid., 22.
47. Ibid., 9.
48. Ibid., 120.
49. Ibid., 101.
50. Ibid., 110.
51. Ibid., 123.
52. Ibid., 124.
53. Ibid., 129.
54. Ibid., 131.
55. Kevin Quashie, *The Sovereignty of Quiet: Beyond Resistance in Black Culture* (New Brunswick: Rutgers University Press, 2012), 22.
56. Jones, *Corregidora*, 102.
57. Recall the moment when a five-year-old Ursa asks, "You telling the truth, Great Gram?" Great Gram slapped her, asserting, "When I'm telling you something, don't you ever ask if I'm lying." In her next sentence, she returns to her compulsory leaving of evidence through storytelling: "Because they didn't want to leave no evidence of what they done – so it couldn't be held against them." Jones, *Corregidora*, 14.
58. As Fiumara notes, "the willingness to keep alive [an] orientation towards openness is the genuine basis for every question. The very notion of question is sustained by an openness – presumably as openness towards listening to the answer." Fiumara, *The Other Side of Language*, 36.
59. Jones, *Corregidora*, 100, 101, 103.
60. Susan Bickford, *The Dissonance of Democracy: Listening, Conflict, and Citizenship* (Ithaca, NY: Cornell University Press, 1996), 2.
61. Danielle Allen, *Talking to Strangers: Anxieties of Citizenship Since Brown v. Board* (Chicago: Chicago University Press, 2006), 45, 46.
62. Dylan Robinson suggests this potential for listening as well: "Developing an awareness of listening positionality here holds potential for listening otherwise, yet the question remains of how – or the extent to which – we might orchestrate such stratified positional listening toward intersectional antiracist, decolonial, queer, and feminist listening practices." Robinson, *Hungry Listening: Resonant Theory for Indigenous Sound Studies* (Minneapolis: University of Minnesota Press, 2020), 60.
63. Fiumara, *The Other Side of Language*, 11, 26.

# Works Cited

Allen, Danielle. *Talking to Strangers: Anxieties of Citizenship since Brown v. Board*. Chicago: University of Chicago Press, 2006.

Bickford, Susan. *The Dissonance of Democracy: Listening, Conflict, and Citizenship*. Ithaca, NY: Cornell University Press, 1996.
Brooks, Daphne A. "Toni Morrison and the Music of Black Life." *Pitchfork*, August 15, 2019. https://pitchfork.com/thepitch/toni-morrison-and-the-music-of-black-life/.
Ceraso, Steph. *Sounding Composition: Multimodal Pedagogies for Embodied Listening*. Pittsburgh: University of Pittsburgh Press, 2018.
Daughtry, J. Martin. "Acoustic Palimpsests and the Politics of Listening." *Music and Politics* 7, no. 1 (2013): 1–34.
Fiumara, Gemma Corradi. *The Other Side of Language: A Philosophy of Listening*. Translated by Charles Lambert. New York and London: Routledge, 1990.
Furlonge, Nicole Brittingham. *Race Sounds: The Art of Listening in African American Literature*. Iowa City: University of Iowa Press, 2018.
Hall, Cheryl. "Beyond the 'Literary Habit': Oral Tradition and Jazz in *Beloved*." *MELUS* 19, no. 1 (1994): 89–95.
Harper, Michael, and Robert B. Stepto, eds. *Chant of Saints: A Gathering of Afro-American Literature, Art, and Scholarship*. Urbana: University of Illinois Press, 1979.
Hartman, Saidiya. *Lose Your Mother: A Journey Along the Atlantic Slave Route*. New York: Farrar, Straus, and Giroux, 2007.
Jones, Gayl. *Corregidora*. Boston, MA: Beacon Press, 1975.
Mackey, Nathaniel. *Discrepant Engagement: Dissonance, Cross-Culturality, and Experimental Writing*. Cambridge: Cambridge University Press, 1993.
Mackey, Nathaniel. *Djbot Baghostus' Run*. Los Angeles: Sun and Moon Press, 1993.
Mackey, Nathaniel. *Paracritical Hinge: Essays, Talks, Notes, Interviews*. Madison: University of Wisconsin Press, 2005.
Moro, Andrea. "What Is the Sound of Thought?" *The MIT Press Reader*, September 18, 2016. https://thereader.mitpress.mit.edu/the-sound-of-thought/.
Morrison, Toni. *Jazz*. New York: Plume, 1993.
Morrison, Toni. *The Dancing Mind*. New York: Knopf, 1996.
Nancy, Jean-Luc. *Listening*. New York: Fordham, 2007.
Quashie, Kevin. *The Sovereignty of Quiet: Beyond Resistance in Black Culture*. New Brunswick: Rutgers University Press, 2012.
Robinson, Dylan. *Hungry Listening: Resonant Theory for Indigenous Sound Studies*. Minneapolis: University of Minnesota Press, 2020.
Rubenstein, Roberta. "Singing the Blues/Reclaiming Jazz: Toni Morrison and Cultural Mourning." *Mosaic: An Interdisciplinary Critical Journal* 31, no. 2 (1998): 147–63.
Scarry, Elaine. *Dreaming by the Book*. Princeton, NJ: Princeton University Press, 2001.
Smith, Bruce R. *The Key of Green: Passion and Perception in Renaissance Culture*. Chicago: University of Chicago Press, 2009.
Smith, Zadie. "Fascinated to Presume: In Defense of Fiction." *The New York Review*, October 24, 2019. https://www.nybooks.com/articles/2019/10/24/zadie-smith-in-defense-of-fiction/.
Stepto, Robert. *A Home Elsewhere: Reading African American Classics in the Age of Obama*. Cambridge, MA: Harvard University Press, 2010.

# 2

# When a Poem "Sounds" through the Body

## Irene Polimante

The relationship between written words and sound is characterized by a long-standing history of debates in literary studies, as well as in postmodern drama, avant-garde theatre, music, folklore studies, among others. In poetry, the ontological and metaphorical implications of "whether texts can really be said to 'speak' [sound] at all"[1] merges with the controversial task to define what the poetic voice is, since this highly ambiguous term encodes a wide "range of aesthetic, cultural, political, and even spiritual attitudes"[2] that challenge the genre itself. For instance, echoing back to avant-garde experimentalism, sound poetry explores new conceptions and uses of sound in poetry to transcend "traditional distinctions between sound and speech, sound and music, sound and noise, music and noise,"[3] and to unveil new meanings by working as a sort of "anti-text" that resists, refracts, and reacts to what the poet rebounds.[4] Moreover, since the 1950s, with the use of tape-recording technologies mixed with experience-based performances, sound poetry has been experiencing a third florescence in Western culture.[5] Inspired by previous experimentations with sound and voice,[6] poets began to use improvisation to mix musical rhythms with their poetry. Previously the poem was "an act consisting of respiratory and auditive combinations, firmly tied to a unit of duration"[7] as well as to the physical presence of the poet, but with the advancements in audio-technology, voice gained its own relevance. Tape recording, cutting, microphones, and speed-change technology provided numerous possibilities for the human voice to reach, once it has been extrapolated from the human body that produced it, and from the "unidirectionality of real-time performance" as well.[8]

Although recording technology seemed to offer a new "prosthetic" for the voice, it did not annihilate the body. On the contrary, it brought new awareness of the importance and potentiality of the physical presence when it concerns the production of vocal sounds, since every voice is related and representative of a specific person. Thus, advancements in recording technology laid the basis for a new understanding of the role of the voice,[9] opening numerous possibilities for poets to craft and present their poems. In this light, the 1950s marked the importance of the aural and performative features of a poem, with sound poetry bridging the gap among written and oral traditions[10] and folklore,[11] becoming a form of popular entertainment especially among marginalized groups in the United States,[12] and leading to the rise of poetry readings for a general dissemination of poetry in North America.[13]

Concerning aurality, poems-in-performance—like poetry readings—highlighted the importance of the audiotext of a poem, the poet's acoustic performance: to wit, "a semantically denser field of linguistic activity than can be charted by means of meter, assonance, alliteration, rhyme, and the like (although these remain underlying elements of this denser linguistic field)."[14] What is heard, therefore, gains as much relevance as what is visually set on page. And to pay attention to the aural features of a poem means to add new layers of meaning, insomuch as the poetic practice acquires more complexity.[15]

The aural dimension of the poem—made of the audiotext as well as of all the other sounds that may be part of the performance—also works as a "junction point" between the linguistic endeavor of the poetic art, and Michael Bakhtin's "bodies of meaning": the nonverbal, embodied dynamics that the textual paradigm is not able to register. Such an experimental effort to combine different textualities (language, audiotext, bodytext) with different modalities (written, oral, digital) is heavily grounded in performance, since performance is considered the ideal field to probe "the limits of intelligibility and referentiality."[16]

But the performative element, with its focus on the direct and/or technologically mediated presence of the body, implies an epistemological rethinking of the poetic genre.[17] This chapter moves from two issues of such a reconceptualization. First, a performative and semiotic understanding of the poem as a dialectic, multilayered process, which is grounded in a complex and dynamic relationship between body, language, and sound. In these terms, the poetic text in all its possible manifestations (written, performed, recorded) needs to be conceived not as a physical reality but as "a concept-limit."[18] Hence, the poem ceases to be a self-contained object,[19] and manifests itself in the communicative (p)act between poet and audience, in which the practice of signification takes place as an inter-human process.[20] This means that the text stays in a continuous tension between the material and the immaterial, opening the theoretical discourse to a post-literary perspective. That is, the poem exists in different states at the same time as intellectual activity, poetic experience, social practice, and aesthetic discourse.[21]

The second point concerns the role of the body and its influence on the poetics practice. The pre-modernist idea of the "semantic body"—a "disciplined, trained and formed" signifier, whose marginalized physicality symbolized "the 'domination of nature applied to the human being' "[22]—has been abandoned in favor of an avant-gardist concept of the body as "*agent provocateur*," which aims not "at the realization of a reality and meaning but at the experience of potentiality."[23] Such a new understanding foregrounds the idea (later explored by post-humanism) of an "anthropological mutation," where the union of man and machine produces "a programmable techno-body," a "controllable and selectable apparatus."[24] One of the numerous implications of this new understanding implies that the body is no more a function (a medium to tell or to represent something else), but instead becomes "a reality of its own" that "*manifests* itself as the site of inscription of collective history."[25]

Moving from these premises, the chapter focuses on three contemporary poets, Rico Frederick, Gabrielle Civil, and Terrance Hayes, who implement different strategies to address the complex relationship among sound, language, and meaning. But they also pay attention to other poetic dimensions, like body and movement, that otherwise tend to be considered collateral to the pervasive debate on the dichotomy between literacy and orality.

## Rico Frederick: Poetic Trans-Positions

Rico Frederick is a Trinidadian graphic designer transplanted in New York, who works between poetry, visual installations, and films to experiment with the miscellaneous convergences of different accents and rhythms produced by the encounter and clash of Creole, African American Vernacular English, and English. Frederick's poetry is heavily grounded on his life spent in-between two different cultures, in the attempt to mediate between the American culture and his Caribbean heritage which sounds through the Caribbean lilt that informs his performances. But poetry is also the form Frederick has chosen to rebuild, elaborate, and retell those legacies that, persisting in a different country, allow the past to coexist with the present through the images, words, and sounds of everyday life. And for this graphic designer and art director, the best way to combine images, languages, and sounds is by crafting the poem as if it was a work of design. In this light, the page embodies the many spaces where imagination can take form. It represents the poet's working space, the place where images, thoughts, words, feelings, and memories may be drawn, becoming the starting point for readers to go wherever they want with their imagination. The blank page is like a drawing sheet. There, the poet translates his imagination in a drawing-like form of writing, where colors, forms, fonts, illustrations, posters, white spaces, black erasures, and even speech bubbles shape English, which sometimes is modulated by the transcription of the Caribbean accent, or abruptly interrupted by the signs of Morse code, when something needs to be said but not immediately seen. This multiform and colorful ensemble of signs creates *Broken Calypsonian* (2014), Frederick's first poetry collection that is set up as the written and drawn version of a Caribbean carnival. The four sections of the volume recall the songs and the days of the Trinidadian celebration ("j'ouvert morning," "Dimanche gras," "road march," and "Calypso monarch"), which also mark the four passages in the poet's journey from Trinidad de Tobago to another island: Manhattan. In this personal march, the poet becomes the *kaiso*, the *chantwell*, the griot of his own life. Each poem, as a fragment of the story, brings light in "that mind tingling bacchanal"[26] where family stories, accounts of childhood and romantic love mingle with issues of race, belonging, masculinity, fatherhood, love, and sorrow:

> Heart / break,
> what drug do I take for the un-nameable pain?
>
> Teething ache the earth cannot swallow,
>     sorrow-clawing at the windowspane,
>   wounded animal[27]

Sorrow and pain are the poet's companions in this existential research, but they are also the propulsive force of his creativity, that feeds itself on the pain, that later will be transformed in poetry: "—I sing of a broken heart,/hurt: is its own form of healing."[28] Thus, accepting the role of a "broken Calypsonian," the poet positions himself in the oral Caribbean tradition:

> Broken Calypso,
>     my heart will sing you for a living.
> The wind in my throat
>         is medicine for steam this dizzyEngine.[29]

But this "singer of truth & folklore" re-visits his role in a contemporary key.[30] And since the visual is the poet's realm, Frederick experiments with poetry, performance, and graphic art, participating in the realization of a 3D film-poem. *HeartBreaker* is "a stereoscopic short film in collaboration with Director Aron Baxter [which] was a 2011 Official Selection of the 8th Annual Big Apple Film Festival."[31] The film is built as a musical video: the background music creates the beat on which the poet reflects on the failures of his romantic life, performing in a hip-hop style the poem written "FOR ANYONE WHO HAS EVER HAD/THEIR HEART BROKEN."[32]

The video begins with a black, silent screen. The two lines just mentioned appear in golden letters, that fluctuate and disassemble to form a spiral which rapidly descend like a swirl to explode at the poet's feet. Sounds effects introduce the two dedicatory sentences and mark every single movement of the letters, while a melody plays in the background. The sound opens the video and also introduces the poet, who starts to speak on the beat that preceded him. As the voice articulates the words, creating the poem, different parts of the setting are projected on a big screen behind him. Whether the voice generates words and meaning, the sound establishes the time, rhythm, movements, and development of such a creation. Slowly, in the background, the urban landscape of a Manhattan by night is digitally reproduced. The sound of the train running, together with other sound effects of the urban soundscape, emphasizes certain passages, disrupting the spoken flow, while the image of a woman dancing at the window of a building behind the poet embodies the poet's ex-girlfriend, for whom "loving me / was the best thing you never did."[33] For the whole video the poet stays almost still at the center of the screen in front of the camera. His action is minimal: The poet mostly uses the upper part of the body (arms, hands, and face) to follow the rhythm of the background music. Changes in shooting—close, medium, and point of view shots alternate frequently—coupled with changes in tone, pitch, and rhythm, mark the pace and give movement to the poem, which opens with an act of creation and ends with an act of disintegration. At the beginning, sound generates words and movement, and the three together form the poet, whose voice creates the poem itself. At the apex of the performance, when the poem reaches its completion, the poet begins to spread golden dust from his hands; the dust turns into golden letters, and the process accelerates involving his whole body. The end of the poem coincides with the end of the poet's voice and the music, as well as the dissolution of the poet's body in thousands of golden letters.

The rhythmic and synchronized intertwining of the poet's movements and words with digital sounds and images allows the fruition of the poem in the form of a music video. Frederick applies the format and technics of the well-known promotional system of the music industry to present his poem. The length of the shooting, the narrative structure, the soundtrack as well as setting, characters, and special effects all concur to the filmic realization of the poem. The strategy at the core of the project seems to aim at two targets: to reach an audience that may be not familiar with the poetic genre, but who is at ease with music; and to experiment with multimodal and multimedia devices to develop the different textualities of the poem—the aural-, the visual-, and the bodytext—together with the work on language. In so doing, Frederick develops a digital version of what Stephen Tyler defined the "new postmodern ethnographic text": "a text of the physical, the spoken, the performed . . . a text to read not with the eyes alone, but with the ears in order to hear 'the voices of the pages.'"[34] A text created by the combination of visual, bodily, aural and language elements.

## Terrance Hayes: America in a Cycle of Sonnets

While Frederick's "video-poem" represents the attempt to act the words, their sounds and meaning through images, giving a sonic and visionary idea of the poet's creative art, Hayes works at a more emotional level. Inspired by the ability of music to communicate bypassing the cognitive processes of understanding, Terrance Hayes plays with form and language to write poems that can "communicate feelings at the base level in the same way a composition with no words communicates meaning."[35] In this light, the poem results from the hard work in unburdening language from the constraints of centuries of theories and cultural superstructures that inform our thinking and expectations. Thus, Hayes's poetics is built to privilege feelings and emotions over language and meaning, in order to create the best conditions for the poet's words to be felt even before being understood.[36]

A sample of such an empathic resonance can be found in Hayes's last collection of poems, *American Sonnets for My Past and Future Assassin* (2018), where Hayes looks at contemporary America through the lens of seventy sonnets that he wrote during the first two hundred days of Trump's presidency. Although the collection is not Trump-centered, Hayes draws on the feelings that pervaded the aftermath of Trump's election as a starting point for a larger discourse. As in a process of psychological introspection, the poet explores his personal darkness, and immerses himself into feelings of anxiety, disillusionment, fear, uneasiness, and hatred. Moving from his own emotional status, Hayes introduces a series of considerations on the past and present of the African American community, on notions of blackness, violence, but also love, childhood, masculinity, and spirituality. The title of the collection is also the title of each single poem, an obsessive repetition that embodies an inner question: whether it is possible to love who or what can kill you,[37] especially when the assassin has hundreds of different colors and shapes but repeats the same old story again and again: "A nigga can survive.

Something happened /. . . In Chicago & Cleveland & Baltimore & happens / Almost everywhere in this country every day."[38]

This passage from personal story to collective history consolidates the long-standing relationship between art and community, while reinforcing an intricate network of memories, experiences, and beliefs that informs a deep sense of collectivity and connectivity within the African American community.[39] Moreover, by directly addressing the assassin(s) with "You," the poet faces brand new forms of old problems, like the Jim Crow laws turned into the more contemporary "gym & crow."[40]

However, in the need for a new perspective to articulate a politically charged poetry, Hayes derives the form of the new American sonnet from both the English and the African American tradition. Concerning the English model, he maintains the traditional fourteen lines, replacing the iambic-pentameter with free verse. Furthermore, he highlights the twist introduced by the last couplet, creating lapidary lines that emphasize the sonic power of the poet's "voltas of acoustics."[41] Whether change is the key element of both Hayes's collection and poetics, it is regulated by a meticulous organization of the poems, which are grouped in five sections of fourteen poems each. Every section is a finite system composed of distinct, self-sufficient poems that work independently from one another. Each poem represents a fragment of Hayes's convoluted journey "to where my blood runs," as the quotation from Wanda Coleman declares in the opening ex-erga.[42] Such a rigid and repetitive organization is pivotal to determining the pace and rhythm of this choral work, where the voice of Hayes intertwines with the many voices that participate in the composition, as Hayes openly addresses other famous authors who have been fundamental in his formation and who also inspire some reflections in the collection. Thus, Hayes introduces the readers to his "hunch," Sylvia Plath, the poet "who does not recognize / Her vision,"[43] and continues with James Baldwin, Jimi Hendrix, Toni Morrison, Emily Dickinson, Prince, who "taught us a real man has / A beautiful woman in him,"[44] and makes indirect references to Yeats. Additionally, there is a continuous passage between real and fictional characters: from George Wallace, Emmett Till, and Martin Luther King, Jr., to Will Smith's character in *Hancock*, Doctor Who and the mythological Eurydice—who, according to Hayes, is the real poet, "not Orpheus. Her muse / Has his back to her with his ear bent to his own heart."[45] This combination of pop culture references with historical and political issues, within a literary framework, provides a polyphonic and multilayered structure, where codeswitching and codemixing go along with a wide use of paronomasia, puns, metaphors, and internal rhythms.

*American Sonnets for My Past and Future Assassin* works as a composition in which the five sections are variations, and the internal repeated lines are refrains that modulate the change in momentum of every single poem.[46] Having abandoned the Shakespearian rhyme scheme, pacing determines the movement of the whole acoustic text, while regulating changes in mood, tension, tone, and speed.

With reference to the African American poetic tradition, Hayes's project is inspired by the work of Wanda Coleman, who proposed a revision "of traditional poetic forms like the sonnet to a range of jazz influences and to the peculiar geographies of neglected

American social strata."[47] But Coleman's *American Sonnets* (1994) was not the first attempt to experiment with one of the most known poetic forms. The relationship between the sonnet and the African American poetic tradition can be traced back to the 1930s, when African American poets used the sonnet as an emancipatory act, a response to "the need for *self-fashioning*," and a "sign of the self-in-process."[48] During those years, indeed, a new understanding of the sonnet took place: from "a privilege-soaked, white-identified form," the sonnet became an element of the Black literary imagination, which informed the formation of "new identities and new psychosocial potentialities."[49] In this light, Hayes's sonnets lay in-between different cultures and times working as junction points of worlds often in contrast. Although the poems are labeled with the same name, they are all as different as the many people, languages, and ethnic groups that constitute the United States. For this reason, Hayes needs a composite and well-organized structure to keep together the many voices and feelings that constitute the *corpus* of his work, where the social, cultural, and historical national body is built through a collage of memories, reflections, emotions, and digressions. While showing the dark sides of the American body, which is, nevertheless, still full of possibilities and capable of resilience, Hayes makes the reader experience what it means to be part of such a vibrant and controversial country.

## Gabrielle Civil: A Trans-media Creative Flow

Gabrielle Civil ends this foray—from Frederick's digital video-poem in hip-hop style to Hayes's complex experimentation in form and pace—with an emphasis on the "corporeal dimension" of poetry.

Gabrielle Civil is an African American performance artist originally from Detroit, who works at the intersection of poetry with conceptual art, and installation. At the core of Civil's poetics there is a strong awareness of her own body: its appearance, shape, color, movements, but also potentialities, and innumerable possibilities of meaning, transformation, communication, and action. As a matter of fact, every single physical aspect concurs in the formation of a corporeal grammar that characterizes both her poetry and performance. Body is the hallmark of Civil's performance art: she enacts the "activation of presence" through her physical presence and, in so doing, the poet performer grounds her creative effort in-between the figurative and the poetic.[50] Moving from the body, Civil investigates how language and voice are deeply connected to her corporeality, highlighting their most material aspect. In this light, performing means to flesh out poetry, to bring poetry to life, "using the body to write into or through spaces, to materialize concepts."[51] Whether the body works as a writing tool, writing becomes "a way of thinking, a way of imagining" what later would be created within the frame, the context, of the performance. In this way, Civil operates an inversion of the common understanding of the relation "writing–performance"—where writing stands for the abstract intellectual activity, and performance refers merely to its material representation.[52] On the contrary, Civil accords equal dignity to the two of them, adopting a more active, intimate, and participated way of knowing—

"knowing how," "knowing who"—which "is anchored in practice and circulated within a performance community."[53]

Furthermore, the relationship among body, writing, and performance shows the many levels at which poetry may operate at the same time, activating multiple meanings, and each time with a different impact or emphasis. Therefore, during a performance, the materiality of the body and the poet's presence in space resonate together with the poet's voice and words bringing the poem to life. Such an emphasis on performance and bodily presence marks the continuity with the African American tradition. In Black culture, performance plays a pivotal role in the process of Black subjectivity formation, since it is historically aligned with issues of hypervisibility, ubiquitously intersected with subjection, and strongly dependent on dynamics of display.[54] However, to Civil, tradition is also the strength and the energy a people derives from their ancestors,[55] the power enabling processes of connectivity and community-bonding. The lineage acknowledged and celebrated by Civil is made of female writers and artists, who contributed to the development of the Black female emancipation by resisting the stereotypes of the "white supremacist capitalist patriarchy," and creating radically different images of the Black woman.[56] In her performance memoire, *Experiments in Joy* (2019), Civil mentions Ntozake Shange, Adrienne Kennedy, Suzanne Césaire, Audre Lorde, Jayne Cortez, Adrian Piper, June Jordan, Octavia Butler, and many other female voices who "don't just rest in power, they radiate it."[57] These Black feminists paved the way for the new feminists of today, who follow in the footsteps of the old "joy" to defy oppression, to oppose the status quo, and "to dream bigger. With satin & song. With flesh & mind & fancy footwork."[58] Hence, the fight against the status quo starts and develops from the body. It is about the body.

The physical presence of the artist, far from being a mere object of the spectator's gaze, turns upside down the relationship of power between who looks and who is looked at. By exposing herself, Civil addresses the audience directly: she provokes the spectators, taking them out of their comfort zone, waiting for whatever response they may or may not give. This is the case of the solo performance "Berlitz" (2004). To address global archetypes and stereotypes in the representation of the Black woman's body, Gabrielle Civil goes back to the old debate about hip-hop and feminism, revisiting Black women's relationship with sexuality, exhibitionism, and money, to understand how "one person's burden become another's pleasure (and vice versa)"; and how this reflection could champion the creation of a "new *lingua franca*" for Black women.[59] The performance was built on the poet reading a letter in Swedish, written for Gabrielle Civil's "Haitian Kreyól" grandmother, as well as on the presentation of an original poem, "Checking Powerful Black Women Writers," delivered in English with a French translation, which was recorded on a tape that played simultaneously with the poet's reading. Moreover, reading and recitation were accompanied by a sequence of actions that should have embodied different figures of Black women: "the black woman carrying the box on her head, the black woman rummaging through the rag pile, the black woman opening a huge birthday surprise full of festive balloons, the black woman throwing down dancing by herself, the black woman scantily clad, gyrating stiffly in a hip hop video."[60] All these characterizations of women represented and questioned different ways in which the Black woman's objectification occurs, following an increasing progression in the intensity of the performance, that reached its apex with Gabrielle Civil's striptease. On high heels and underwear, she "pretended to be vacant, screwed [her] finger into [her] cheek, gyrated" and put herself into a carton box.[61] The playful over-exhibition of her body was meant to "contrast the way a black woman's body could feel to herself, dancing joyfully without many clothes, to the way a black woman could look objectified in the same attire in a different (global) context."[62]

Gabrielle Civil foregrounds her own physicality in what she labels "fat Black performance art": a form of expressions that provides the poet with "the opportunity to shift the balance between what is expected of me—as a plump, dark-skinned, natural-haired, Black woman—and what I will actually do."[63] Performance and poetry open spaces "to retwist the locks," to enact an "erotic scholarship" that "entails speaking from, for, and about the body."[64] In this context, hissing, crawling, and confronting are ways to provoke and subvert established relations of power, like in the performance "Touch/Don't Touch," where the poet deals with the "hair discourse." Hair texture, like skin tone, is "central to understanding the Black female experiences of hair/body politics," since they are historically, culturally, and aesthetically at the basis of the African American experience of discrimination. Concerning African American women, hair embodies their struggle in defining their identity and their relationship to their communities and to the dominant culture as well.[65]

Civil's performance "Touch/Don't Touch" derives from the gauche and sloppy white behavior to touch a Black person's hair without permission, "and even to ask to touch it," since the act of touching recalls slavery discourses on Black inhumanity and inferiority.[66] During the performance she sticks her hair "in the face of a smiling, bearded white man in glasses," who sits on the floor, asking him to touch her hair.[67] The more uneasy the man becomes, the more she bends and leans over him, insisting: "*You know you want to . . .*" and again "*Just doooooo it. Just dooooooooo it.*"[68] And when finally the man taps her "enticing, fuzzy crown" she immediately reacts: "*HOW DARE YOU!*" and, standing on her feet away from him she continues, "*YOU TOUCHED MY HAIR?!*"[69] The interaction created by Civil blurs "the space between humor and threat"[70] to renegotiate a new space of action for the Black woman, who defines a new behavioral pattern, controlling and leading the new relationship of power with the white man. It is the woman who is in charge, "bossing, demanding, cajoling, changing [her] mind," putting the man in an uncomfortable position where he can only wait for the woman's next move. Civil thus emphasizes some of the elements informing the racist representation of the Black woman to turn them into elements of power. In Civil's performance, the body "speaks" its truth, interacting with language, space, and other bodies that also tell of their experiences. Thus, the physical and personal experience not only offers material for the poem or the performance, but it is key to entering into contact with other physical and personal, or collective, experiences. The personal experience, thus, becomes the starting point for an intense dialogue that Civil has with other women artists, in order to address issues of blackness and sisterhood within a fluid and global contemporaneity.

## Sounds and Bodies

Such a brief analysis has shown how these three poets implement different strategies to perform, both on and off the page, an ensemble of different kinds of texts that together constitute what we call a poem. This means that the written page, the audiotext or "the audible acoustic text of the poem,"[71] as well as the *bodytext* or the physical, tactile-kinesthetic text of the poem, are the foundations of the poetic praxis. Their role and influence in poets' poetic practice are visible thanks to the lens of performance. As a matter of fact, the physical presence of the body, coupled with performative elements, is more visible in performance poetry.[72]

This chapter has provocatively addressed many debated issues in literary criticism to boost a more holistic approach to the study of the relations between sound/voice, words, and body. It is critical that we open our understanding of poetry in order to consider the multiple potentialities and possibilities of a genre that is something more than a literary activity, that is a personal experience, a social practice, and an aesthetic discourse, which are not only built on words.

## Notes

1. Richard Aczel, "Hearing Voices in Narrative Texts," *New Literary History* 29, no. 3 (1998): 467.
2. Lesley Wheeler, *Voicing American Poetry: Sound and Performance from the 1920s to the Present* (Ithaca, NY: Cornell University Press, 2008), 2.
3. Nacy Perloff, "Sound Poetry and the Musical Avant-Garde: A Musicologist's Perspective," in *The Sound of Poetry/The Poetry of Sound*, ed. Marjorie Perloff and Craig Dworkin (Chicago: University of Chicago Press, 2009), 97.
4. Steve McCaffery and bpNichol, *Sound Poetry: A Catalogue for the Eleventh International Sound Poetry Festival, Toronto, Canada, October 14 to 21, 1978* (Toronto: Underwhich Editions, 1978).
5. Steve McCaffery, "The Emergence of the Audio-Poem," in *Sound States: Innovative Poetics and Acoustical Technologies*, ed. Adelaide Morris (Chapel Hill: The University of North Carolina Press, 1997), 146–68.
6. I'm referring, for example, to the work of Dadaist sonosophers as well as the phonetic performances of twentieth-century European and Russian avant-gardes; the use of tautograms and call and response, like in Cab Calloway's poems; Henry Chopin's audio-poems; and the complex framework of bluesy sounds and rhythm in Langston Hughes's poetry.
7. McCaffery, "The Emergence of the Audio-Poem," 153.
8. Ibid., 157.
9. Simon Frith detects four functions of voice in popular music: voice works as a musical instrument; as a body, revealing acoustic features that are easily decodifiable in terms of gender, age, or ethnicity; as a person, becoming the vocal expression of a particular individuality (with a uniqueness similar to fingerprints); and as a character, when who is performing "put[s] on a vocal costume, enacting the role that they are playing for ourselves" (Frith, "The Voice," in *Performing Rites: On the Value*

of *Popular Music*, ed. Simon Frith [Cambridge, MA: Harvard University Press, 1998], 183–202). Notwithstanding the differences between singers and poets—especially in terms of "performing a character"—these four characteristics can nevertheless be traced in any kind of poetic performance as well.

10   In relation to the long-standing debate on the so-called "Great Divide" between literacy and orality, see: Paul Zumthor, *Oral Poetry: An Introduction* (Minneapolis: University of Minnesota Press, 1990); Gregory Nagy, *Poetry as Performance: Homer and Beyond* (Cambridge: Cambridge University Press, 1996); Egbert J. Bakker, *Poetry in Speech: Orality and Homeric Discourse* (Ithaca: Cornell University Press, 1997); and Walter J. Ong, *Orality and Literacy: The Technologizing of the Word* (New York: Routledge, 2002).

11   S. J. Sackett lists "some points of affinity" between poetry and folklore, providing a few examples on the use of simile, metaphor, rhyme, assonance, alliteration, assonance, rhythm, and so on. See: Sacket, "Poetry and Folklore: Some Points of Affinity," *The Journal of American Folklore* 77, no. 304 (1964): 143–53.

12   Nonverbal and extralinguistic forms of communication become means of subversion for oppressed and subaltern people that, excluded from literacy, develop another kind of hermeneutic, in which knowledge is not detached and abstracted, but rather develops as a "located activity" (Conquergood, "Beyond the Text: Toward a Performative Cultural Politics," in *The Future of Performance Studies: Visions and Revisions*, ed. Sheron J. Dailey [Washington, DC: National Communication Association, 1998], 27). In opposition to an idea of objectivity, "[p]roximity, instead of purity, becomes the epistemological point of departure and return" (Dwight Conquergood, "Performance Studies," *TDR* 46, no. 2 (2002): 28). Such a subversion of the Western "textual attitude" (Said, *Orientalism* [London: Penguin, 1979], 368) becomes crucial thanks to the performative turn. In poetry, nonverbal and extralinguistic communication has gained more relevance with the spread of slam and spoken word poetry since the late 1980s.

13   Charles Bernstein, introduction to *Close Listening: Poetry and the Performed Word*, ed. Charles Bernstein (Oxford: Oxford University Press, 1998), 5.

14   Ibid., 13.

15   In those years the Beats were a good example. But, in time, more and more poets and artists paid particular attention to the power of voice, rhythm, and the aural experience a poem can give. I am referring to Amiri Baraka, Bob Kaufman, Cecil Taylor, John Cage, and John Giorno (just to mention a few among many).

16   Perloff, "Sound Poetry and the Musical Avant-Garde," 98.

17   In recent years a few researchers have seriously investigated the topic, like Martina Pfeiler, *Sounds of Poetry: Contemporary American Performance Poets* (Tübingen: Narr Verlag, 2003); Peter Middleton, *Distant Reading: Performance, Readership, and Consumption in Contemporary Poetry* (Tuscaloosa: The University of Alabama Press, 2005); Julia Novak, *Live Poetry: An Integrated Approach to Poetry in Performance* (Amsterdam and New York: Rodopi, 2011); and Birgit M. Bauridl, *Betwixt, Between, or Beyond? Negotiating Transformations from the Liminal Sphere of Contemporary Black Performance Poetry* (Heidelberg: Universitätsverlag Winter, 2013), among others.

18   Cesare Segre, "Testo," in *Enciclopedia Einaudi* (Torino: Einaudi, 1981), 269–91.

19   See: Katherine N. Hayles, "The Time of Digital Poetry: From Object to Event," in *New Media Poetics: Contexts, Technotexts, and Theories*, ed. Adalaide Morris

(Cambridge, MA: MIT Press, 2006), 181–210; Maria Damon, *Postliterary America: From Bagel Shop Jazz to Micropoetries* (Iowa City: University of Iowa Press, 2011).

20  See: Cesare Segre, *Ritorno alla critica* (Torino: Einaudi, 2001); Raul Mordenti, *L'altra critica. La nuova critica della letteratura tra studi culturali, didattica e informatica* (Roma: Meltemi, 2007).
21  Damon, *Postliterary America*, 191–202.
22  Hans-Thies Lehmann, *Postdramatic Theatre* (New York: Routledge, 2006), 162.
23  Ibid., 163.
24  Ibid., 165.
25  Ibid., 97, 163.
26  Rico Frederick, *Broken Calypsonian* (New York: Penmanship Books, 2014), 10.
27  Ibid.
28  Ibid., n.p.
29  Ibid., 10.
30  Ibid., n.p.
31  Rico Frederick, "Poetry," last modified October 18, 2020, http://ricofdk.squarespace.com/poetry-1.
32  Rico Frederick, "Heartbreaker Film," directed by Aron Baxter, released 2011, video, 5:03, https://ricofdk.squarespace.com/video.
33  From the video of the poem. For the whole performance, as well as the videos recorded by the poet during the video shoot, see: Frederick, "Heartbreaker Film."
34  Stephen A. Tyler, "Post-Modern Ethnography: From Document of the Occult to Occult Document," in *Writing Culture: The Poetics and Politics of Ethnography*, ed. James Cliffords and George E. Marcus (Berkeley: University of California Press, 1986), 136.
35  Terrance Hayes, "Terrance Hayes," *Poetry Foundation*, accessed October 18, 2020, https://www.poetryfoundation.org/poets/terrance-hayes.
36  Ibid.
37  "I think this dude is trying to kill me . . . and can I still love him? Can I write a sonnet for my assassin?" See: Don Share, Lindsay Garbutt, and Terrance Hayes, "Terrance Hayes Reads 'American Sonnet for My Past and Future Assassin,'" September 11, 2017, in *The Poetry Magazine Podcast*, MP3 audio, 11:16, https://www.poetryfoundation.org/podcasts/144101/terrance-hayes-reads-american-sonnet-for-my-past-and-future-assassin.
38  Terrance Hayes, *American Sonnets for My Past and Future Assassin* (London: Penguin, 2018), 9.
39  Meta DuEwa Jones, *The Muse Is Music: Jazz Poetry from the Harlem Renaissance to Spoken Word* (Chicago: University of Illinois Press, 2013).
40  Hayes, *American Sonnets for My Past and Future Assassin*, 11.
41  Hayes, "Terrance Hayes,".
42  Hayes, *American Sonnets for My Past and Future Assassin*, n.p.
43  Ibid., 5.
44  Ibid., 63.
45  Ibid., 59.
46  Metta Sáma, "The Burden of Seed, the Seed of Burden: 'Repetitional Schemas' & Pace in Terrance Hayes's 'Sonnet,'" in *Mentor and Muse: Essays from Poets to Poets*, ed. Blas Falconer, Beth Martinelli, and Helena Mesa (Carbondale and Edwardsville: Southern Illinois University Press, 2018), 101–6.

47  Jennifer D. Ryan, *Post-Jazz Poetics: A Social History* (New York: Palgrave Macmillan, 2010), 16.
48  Jon Woodson, *Anthem, Sonnets, and Chants: Recovering the African American Poetry of the 1930s* (Columbus: The Ohio State University Press, 2011), 11.
49  Ibid., 12.
50  Ilana Masad and Gabrielle Civil, "Gabrielle Civil's Art Is an Experiment in Joy," *Nylon*, September 10, 2019, https://nylon.com/gabrielle-civil-experiments-in-joy.
51  Ibid.
52  In Western culture, writing embodies the dominant distanced perspective of a system of knowledge-formation based on the purpose of "knowing that" and "knowing about," insomuch as it describes "a view from the above of the object of inquiry: knowledge that is anchored in paradigm and secured in print" (Conquergood, "Performance Studies," 145).
53  Ibid.
54  See: Cedric J. Robinson, *Black Marxism: The Making of the Black Radical Tradition* (Chapel Hill: The University of North Carolina Press, 1983); Soyica Diggs Colbert, *The African American Theatrical Body: Reception, Performance, and the Stage* (Cambridge: Cambridge University Press, 2011). See also Fred Moten's theorization of the "phonic substance" that can be heard in the acoustic space opened by the performance (Fred Moten, *In the Break: The Aesthetics of the Black Radical Tradition* [Minneapolis: The University of Minnesota Press, 2003]).
55  Gabrielle Civil, *Experiment in Joy* (New York: The Accomplices, 2018).
56  bell hooks, *Sisters of the Yam: Black Women and Self-Recovery* (New York: Routledge, 2015).
57  Gabrielle Civil, *Experiment in Joy*, 13.
58  Ibid.
59  Gabrielle Civil, *Swallow the Fish: A Memoir in Performance Art* (New York: Civil Copying Mechanism and the Accomplices, 2017), 258.
60  Ibid.
61  Ibid.
62  This contrast was even more amplified by a general rate of dissatisfaction that almost 96% of American women have with their bodies. As reported by psychologist Linda Siemanski's account of such a social plague to the poet, "[w]hether they're thin or fat, tall or short—their normal state is to believe that something is wrong with them. In fact, a mark of assimilation for immigrant women is their rising rate of body dissatisfaction" (Civil, *Swallow the Fish*, 258).
63  Gabrielle Civil, "Touch/Don't Touch," *Art21 Magazine*, April 19, 2017, http://magazine.art21.org/2017/04/19/touchdont-touch/#.XmgSUpNKi5w.
64  Civil, *Swallow the Fish*, 111.
65  On the topic, see: bell hooks, *Black Looks: Race and Representation* (New York: Routledge, 1992); Ingrid Banks, *Hair Matters: Beauty, Power, and Black Women's Consciousness* (New York: New York University Press, 2000); and Robin D. G. Kelly, "Nap Time: Historicizing the Afro," *Fashion Theory* 1, no. 4 (1997): 339–52.
66  Civil, "Touch/Don't Touch," n.p.
67  Ibid.
68  Ibid.
69  Ibid.

70 Bernstein, introduction to *Close Listening*, 5.
71 Ibid., 12.
72 The peculiar characteristic of performance poetry is to work among multiple media in the attempt to incorporate sounds, body movements, facial expressions, communication skills, language(s), and, sometimes, technological devices too.

## Works Cited

Aczel, Richard. "Hearing Voices in Narrative Texts." *New Literary History* 29, no. 3 (1998): 467–500.
Bakker, Egbert J. *Poetry in Speech: Orality and Homeric Discourse*. Ithaca, NY: Cornell University Press, 1997.
Banks, Ingrid. *Hair Matters: Beauty, Power, and Black Women's Consciousness*. New York: New York University Press, 2013.
Bauridl, Birgit M. *Betwixt, Between, or Beyond? Negotiating Transformations from the Liminal Sphere of Contemporary Black Performance Poetry*. Heidelberg: Universitätsverlag Winter, 2013.
Bernstein, Charles. "Introduction." In *Close Listening: Poetry and the Performed Word*, edited by Charles Bernstein, 3–26. New York and Oxford: Oxford University Press, 1998.
Civil, Gabrielle. *Swallow the Fish: A Memoir in Performance Art*. New York: Civil Copying Mechanism and The Accomplices, 2017.
Civil, Gabrielle. "Touch/Don't Touch." *Art21 Magazine*, April 19, 2017. http://magazine.art21.org/2017/04/19/touchdont-touch/#.XmgSUpNKi5w.
Civil, Gabrielle. *Experiment in Joy*. New York: The Accomplices, 2019.
Colbert, Soyica Diggs. *The African American Theatrical Body: Reception, Performance, and the Stage*. Cambridge: Cambridge University Press, 2011.
Conquergood, Dwight. "Beyond the Text: Toward a Performative Cultural Politics." In *The Future of Performance Studies: Visions and Revisions*, edited by Sheron J. Dailey, 25–36. Washington, DC: National Communication Association, 1998.
Conquergood, Dwight. "Performance Studies: Interventions and Radical Research." *TDR* 46, no. 2 (2002): 145–56.
Coupland, Nikolas. "Voice, Place and Genre in Popular Song Performance." *Journal of Sociolinguistics* 15, no. 5 (2011): 573–602.
Damon, Maria. *Postliterary America: From Bagel Shop Jazz to Micropoetries*. Iowa City: University of Iowa Press, 2011.
Frederick, Rico. "Heartbreaker Film." Directed by Aron Baxter. Released 2011. Video, 5:03. https://ricofdk.squarespace.com/video.
Frederick, Rico. *Broken Calypsonian*. Brooklyn: Penmanship Books, 2014.
Frederick, Rico. "Poetry." Last modified October 18, 2020. http://ricofdk.squarespace.com/poetry-1.
Frith, Simon. "The Voice." In *Performing Rites: On the Value of Popular Music*, 183–202. Cambridge, MA: Harvard University Press, 1998.
Hayes, Terrance. *American Sonnets for My Past and Future Assassin*. London: Penguin, 2018.
Hayes, Terrance. "Terrance Hayes." *Poetry Foundation*. Accessed October 18, 2020. https://www.poetryfoundation.org/poets/terrance-hayes.

Hayles, N. Katherine. "The Time of Digital Poetry: From Object to Event." In *New Media Poetics: Contexts, Technotexts, and Theories*, edited by Thomas Swiss and Adalaide Morris, 181–210. Cambridge, MA: MIT Press, 2006.
hooks, bell. *Black Looks: Race and Representation*. London and New York: Routledge, 1992.
hooks, bell. *Sisters of the Yam: Black Women and Self-Recovery*. New York and London: Routledge, 2015.
Ilana, Masad, and Gabrielle Civil. "Gabrielle Civil's Art Is an Experiment in Joy." *Nylon*, September 10, 2019. https://nylon.com/gabrielle-civil-experiments-in-joy.
Jenkins, Tammie. *A Case Study of Tracie Morris' Project Princess*. Baton Rouge: Louisiana State University, 2013.
Jones, Meta DuEwa. *The Muse Is Music: Jazz Poetry from the Harlem Renaissance to Spoken Word*. Urbana, Chicago, and Springfield: University of Illinois Press, 2013.
Kelley, Robin D. G. "Nap Time: Historicizing the Afro." *Fashion Theory* 1, no. 4 (1997): 339–52.
Lehmann, Hans-Thies. *Postdramatic Theatre*. London and New York: Routledge, 2006.
McCaffery, Steve. "The Emergence of the Audio-Poem." In *Sound States: Innovative Poetics and Acoustical Technologies*, edited by Adelaide Morris, 149–68. Chapel Hill: The University of North Carolina Press, 1997.
McCaffery, Steve, and bpNichol. *Sound Poetry: A Catalogue for the Eleventh International Sound Poetry Festival, Toronto, Canada, October 14 to 21, 1978*. Toronto: Underwhich Editions, 1978.
Middleton, Peter. *Distant Reading: Performance, Readership, and Consumption in Contemporary Poetry*. Tuscaloosa: The University of Alabama Press, 2005.
Mordenti, Raul. *L'altra critica: La nuova critica della letteratura tra studi culturali, didattica e informatica*. Roma: Meltemi, 2007.
Moten, Fred. *In the Break: The Aesthetics of the Black Radical Tradition*. Minneapolis: The University of Minnesota Press, 2003.
Nagy, Gregory. *Poetry as Performance: Homer and Beyond*. Cambridge: Cambridge University Press, 1996.
Novak, Julia. *Live Poetry: An Integrated Approach to Poetry in Performance*. Amsterdam and New York: Rodopi, 2011.
Ong, Walter J. *Orality and Literacy: The Technologizing of the Word*. London and New York: Routledge, 2002.
Perloff, Nancy. "Sound Poetry and the Musical Avant-Garde: A Musicologist's Perspective." In *The Sound of Poetry/The Poetry of Sound*, edited by Marjorie Perloff and Craig Dworkin, 97–117. Chicago and New York: University of Chicago Press, 2009.
Pfeiler, Martina. *Sounds of Poetry: Contemporary American Performance Poets*. Tübingen: Narr Verlag, 2003.
Robinson, Cedric J. *Black Marxism: The Making of the Black Radical Tradition*. Chapel Hill: The University of North Carolina Press, 2000.
Ryan, Jennifer D. *Post-Jazz Poetics: A Social History*. New York: Palgrave Macmillan, 2010.
Sackett, Sam J. "Poetry and Folklore: Some Points of Affinity." *The Journal of American Folklore* 77, no. 304 (1964): 143–53.
Said, Edward W. *Orientalism*. London: Penguin, 1979.
Sáma, Metta. "The Burden of Seed, the Seed of Burden: 'Repetitional Schemas' & Pace in Terrance Hayes's 'Sonnet.'" In *Mentor and Muse: Essays from Poets to Poets*, edited by

Blas Falconer, Beth Martinelli, and Helena Mesa, 101–6. Carbondale and Edwardsville: Southern Illinois University Press, 2010.

Segre, Cesare. "Testo." In *Enciclopedia Einaudi*, 269–91. Torino: Einaudi, 1981.

Segre, Cesare. *Avviamento all'analisi del testo letterario*. Torino: Einaudi, 1999.

Segre, Cesare. *Ritorno alla critica*. Torino: Einaudi, 2001.

Share, Don, Lindsay Garbutt, and Terrance Hayes. "Terrance Hayes Reads 'American Sonnet for My Past and Future Assassin.'" *The Poetry Magazine Podcast*. September 11, 2017. Podcast, MP3 audio, 11:16. https://www.poetryfoundation.org/podcasts/144101/terrance-hayes-reads-american-sonnet-for-my-past-and-future-assassin.

Spellers, Regina E. "The Kink Factor: A Womanist Discourse Analysis of African American Mother/Daughter Perspectives on Negotiating Black Hair/Body Politics." In *Understanding African American Rhetoric: Classical Origins to Contemporary Innovations*, edited by Ronald L. Jackson II and Elaine B. Richardson, 223–43. New York and London: Routledge, 2003.

Tyler, Stephen A. "Post-Modern Ethnography: From Document of the Occult to Occult Document." In *Writing Culture: The Poetics and Politics of Ethnography*, edited by James Clifford and George E. Marcus, 122–40. Berkeley, Los Angeles, and London: University of California Press, 1986.

Wheeler, Lesley. *Voicing American Poetry: Sound and Performance from the 1920s to the Present*. Ithaca, NY: Cornell University Press, 2008.

Woodson, Jon. *Anthem, Sonnets, and Chants: Recovering the African American Poetry of the 1930s*. Columbus: The Ohio State University Press, 2011.

Zumthor, Paul. *Oral Poetry: An Introduction*. Minneapolis: University of Minnesota Press, 1990.

# 3

# Practices of Unmixing

## Film Aesthetics, Sound, and the New Hollywood Cinema

Christof Decker

After the release of *Nashville* in 1975, film director Robert Altman was interviewed about the sound system that had been prominently cited in the credit sequence as "lion's gate 8 track sound." Altman had formed the production company Lion's Gate in the early 1970s while "8 track sound" referred to technical equipment allowing the film team to record numerous sound sources simultaneously on separate tracks. It was the same technology then being used for the recording of music, as Altman explained, but contrary to the suggestions by some of the interviewers, it did not necessarily create a more realistic sound. "[I]t's really unmixing rather than mixing sound," he said.[1] Still it helped to establish his signature practice of recording various sources and sounds separately, which, under traditional circumstances, would have created one recording of overlapping sonic elements. It therefore allowed Altman to "unmix" the flow of live voices and sounds in order to rearrange and remix them during the process of postproduction in new and surprising constellations.

In this way, Altman's films from the early 1970s participated in auditory practices of the New Hollywood Cinema that had introduced densely layered, irritating, painful, or, more generally, disruptive forms of sound contributing to an oppositional aesthetic. In an institutional Hollywood system destabilized economically during the 1960s and challenged artistically by the innovative European art cinema as well as the domestic schools of documentary and avant-garde, these practices were shaped by subverting the traditions of Hollywood cinema but also by experimenting with technology, ultimately creating new forms of narration and subjectivity. Sound as a disruptive element thus had different connotations. In general terms, by going against audience expectations, it contributed to the New Hollywood Cinema's critique of genre as ideology and popular mythology. More specifically, its subversive character became manifest in the introduction of new soundscapes—such as popular or electronic music—and by transforming traditional forms of mixing. This included a softening of the vertical mode of mixing that centered on the audibility of a limited number of "heroic" characters, and led to approaches aiming to establish a horizontal mode by bringing together in more equal ratios the voices of central and marginal characters as well as the sounds of off-screen and on-screen spaces. In order to frame my analysis of the New Hollywood Cinema's historical moment, I will begin by discussing the place of sound in film studies and, more specifically, in the discourse about film aesthetics. Although this question would deserve a most comprehensive treatment, I will focus on the discourse about the functions of sound, its hybrid character, and the notion of film sound as a stylistic element before returning to Robert Altman's *Nashville*.

## Film Studies and the Question of Sound

A common topic of much recent work in auditory or sound studies has been the observation that sound as an object of study has been neglected for too long. While the visual in visual studies had often found itself to be regarded as inferior in relation to language, the auditory sees itself as being placed in an inferior position vis-à-vis the visual. Following Jonathan Sterne, sound is often stereotyped as being less objective, less active, or less intellectual than the visual, thereby implicitly creating a cultural hierarchy of language at the top, the visual in the middle, and the auditory at the bottom.[2] For the question of placing sound in film studies, this points to an initial paradox. Although the development of sound technology, the uses of sound, and its myriad narrative potentials have a long and complex history dating back to the so-called silent period, its neglect in historical studies is striking. It is no exaggeration to say that in established approaches such as genre and auteur studies, interpretations discussing race, gender, or class, even in industrial or institutional histories, the qualities and significance of sound have usually been secondary to questions of visuality, or they have been discussed as a minor aspect of film music. This is true for the study of fiction films from mainstream Hollywood or independents but also for avant-garde cinema and the history of documentary film, where, incidentally, much innovative work with sound originated from the 1930s onward. Film scholar Bill Nichols published a seminal essay in 1983 called "The Voice of Documentary," yet studies focusing on the auditory quality of non-fiction filmmaking as an integral part of film aesthetics are rare.[3]

However, a growing number of publications aim to rectify this imbalance. While film music in particular has produced a substantial research tradition, sounds coded as non-musical or non-speech-related are certainly prime candidates for a more thorough interest in their changing meanings and historical significance.[4] Writing about sound design, film scholar Helen Hanson points out that "the long and complex history of this crucial, yet largely invisible work and the technicians who executed it still remain largely unknown in film history."[5] But as her study of the classical era demonstrates so vividly, key aspects of the history of sound are finally being investigated at both the macro and micro levels. Scholars are exploring the different stages of recording, creating, mixing, and using sound across the industry but also zooming in on "the activity of the sound craft" for individual films or studios.[6] This shift toward hitherto neglected areas in recent work by, among many others, Rick Altman, Helen Hanson, and Jay Beck, has opened up exciting fields of scholarly research that aim to integrate the study of sound into sophisticated networks of institutional, technological, and narrational histories.[7]

An important publication pointing the way at the level of introductory textbooks is *Hearing the Movies* by James Buhler and David Neumeyer from 2010. However, even this publication, written by two musicologists, prioritizes film music as the prime element on the soundtrack. The authors distinguish between speech, music, and sound effects.[8] Yet ironically, in a publication intent on "hearing" the movies, not all sonic elements are equal. By default, it seems, speech is seen to be most important for providing story information. Music establishes the terms and categories as well as the cultural prestige for the analysis of sonic qualities. And everything else is subsumed under the label of "sound effects" or "noise." While the term "sound effects" seems to imply the predominance of imitation and artificiality, "noise" usually references unwanted or meaningless sounds. Clearly, this terminology is inadequate—not just for soundtracks in the digital age—and more sophisticated concepts are needed for non-musical and non-speech-related sounds.

In film aesthetics the place of sound has traditionally been determined at a functional level. Phenomenologically speaking, sounds are regarded to have material qualities such as loudness, pitch, or timbre based on the amplitude and frequency of their vibrations as well as their tone or "color." By interacting with the visual track, sounds create rhythmic structures. They relate to sources that may be on- or offscreen, diegetic or nondiegetic, and they establish basic temporal and spatial dimensions. Sound gives volume to space, creates proximity or distance, and it establishes internal or external realms. Moreover, it is crucially responsible for the impression of synchronicity and the simultaneous or non-simultaneous relationship with the image.[9] Thus it is obvious that, by making vital contributions to the time, space, progression, and meaning of fictional as well as factual narratives, the properties of sound represent an integral aspect of film aesthetics.

Following Mervyn Cooke and others, sounds and music are seen to have certain functions that underline their centrality to the system of narration. They may suggest "atmosphere, emotions, character traits and specific period or locational settings."[10] They can provide unity or rupture, continuity or rhythmic contrast. They use techniques such as underscoring, mickey-mousing, or counterpoint to support, mimic, or contrast actions and body movements. And, finally, they may strengthen weak scenes, manipulate space and depth, and are integrated into the basic dynamic of foreground and background at the heart of audiovisual storytelling.[11] And yet, even though the rich discourse about the properties and functions of sound can be traced back to the 1930s, where it developed from the perspectives of film craft and film aesthetics, its current place in the field of film studies is contested. This is made clear in a 2014 theoretical overview of film music and sound by James Buhler, and published in *The Oxford Handbook of Film Music Studies*, in which he very usefully sketches some of the major theories dealing with sound and music.[12] However, he also replays the theory wars from the 1990s by contrasting the neo-formalism of authors such as David Bordwell, Kristin Thompson, and Noël Carroll with critical theories put forward by Kaja Silverman, Stephen Neale, and others.

For film studies today, pitting neo-formalism against critical theories in this way seems to be less helpful and ultimately inadequate. To be sure, most scholars accept the premise that sounds have properties and functions in the design of aesthetic objects such as films. In order to evaluate these functions, then, the focus must ultimately shift to the historical norms and conventions regulating the uses of sounds. This is also a basic idea in Buhler's essay. Yet he goes on to argue that for neo-formalist authors historical norms are "relatively neutral stylistic markers that serve above all to organize technique," while critical theories, on the other hand, see in "style and technique the marks of deep cultural ideology."[13] Thus, according to Buhler, neo-formalist theories understand norms to be neutral while critical theories see them as ideological. Both neo-formalist and critical theories, therefore, acknowledge the central importance of historical norms, but they approach them differently. According to Buhler, neo-formalism focuses on aesthetics, formal principles, and matters of style, while critical theories, beginning with *Composing for the Films* in 1947 by Theodor W. Adorno and Hanns Eisler, foreground the activity of interpretation and presuppose that the aesthetic is overdetermined by ideology. Consequently, critical theories explore how sound and music are embedded in ideologies that shape, for instance, ethnic and gender stereotypes or colonial encounters.

Although Buhler's review of individual approaches to the study of sound is on the whole very useful, film aesthetics and ideology critique do not have to be viewed as mutually exclusive. To be sure, neo-formalist theories have often focused on functions, asking, for example, how certain formal elements are related to the design of the whole.[14] Yet, ideology critique has also often used the term function to make a point about the workings of ideology. While neo-formalism usually discusses the *internal* functions of elements making up the film and its narrative, ideology critique often starts with *external* functions of an ideological system, such as the idea of heteronormativity, and then works its way to the individual film where the ideological struggles are presumed to be playing out. For the study of the soundtrack in film history, therefore, we need a functional model which addresses, on the one hand, the material qualities of sounds in relation to film design and film aesthetics and, on the other, the idea that norms must be related to larger ideological forces at work in a particular culture. However, instead of contrasting formalist versus ideological theories of functions, it is more productive to consider three interrelated levels connecting design functions and their *potential* effects with an intermediary institutional level and, furthermore, with cultural functions and *actual* historical uses.[15]

To give an example, Fred Zinnemann's classic Western *High Noon* includes an intensely dramatic scene before the climactic shoot-out: Marshal Will Kane (Gary Cooper) begins to write his last will as the clock starts moving toward twelve o'clock, the arrival time of his antagonist. The score by Dimitri Tiomkin picks up the movement of the clock, while close-ups show the frightened inhabitants of the town anticipating the deadly encounter. Finally, after the music has been building up to a climax, it stops when the sharp, drawn-out whistle of the approaching noon train is heard breaking through the atmosphere of waiting and dread. Faces look up and bodies tighten as the train's whistle pierces through the minds of the characters and the film audience alike to announce the beginning of the show-down. From a design perspective, the sound of the train whistle is a rare instance of non-musical sound serving as a focal point of highest dramatic intensity. However, at the other end of the model, in the realm of cultural and symbolic functions, it is equally interesting as the train whistle signifies the Western genre's clash of technology and "uncivilized" space, bringing together narrative intensity with genre mythology in exemplary fashion. In short, instead of conceptualizing the functions of sound in mutually exclusive formalist or ideological terms, a multilayered model would be a more flexible alternative enabling us to investigate how aesthetic norms may be related to the historical workings of ideological norms.

## The Style(s) of Sound

In his critique of neo-formalism, Buhler claims that the formalist is primarily interested "in what the soundtrack does, not what a particular sound is."[16] This distinction between the purposes and the qualities of sounds, between sound as merely a rhetorical device versus an object with material and medial characteristics, allows us to shift the discussion to another important topic of aesthetics, the question of style. Buhler's claim that the study of sound in film should pay more attention to its material qualities echoes broader calls in sound studies for a shift toward sonic imaginations and the practices of reduced or deep listening. As Michael Bull and Les Beck write, this means learning to listen properly and, in particular, paying close attention to the "multiple layers of meaning potentially embedded in the same sound."[17] For film studies, this ideal of deep listening revisits some old questions, in particular the degree of autonomy we attribute to individual elements making up the cinematic experience. If we agree that non-musical and non-speech-related sounds have been neglected in historical studies, does that mean that we claim a special status for them on the soundtrack? Put differently, does it make sense to practice deep listening with films?

In some respects, if it helps to train the listening skills of the viewers, deep listening may be a valuable practice. But as Adorno and Eisler argued at great length in the 1940s about the case of music, the hybrid character of film complicates its status. For the two authors, film music was fundamentally different from autonomous music and they claimed that it should maintain a sense of detachment and reflexivity vis-à-vis its place and purpose.[18] Put differently, film music should not take itself too seriously because it is always perceived in relation to the image. Indeed, for Adorno and Eisler this relationship between sound and image was a major concern and much of their critique addressed the fact that American cinema usually saw music as having to serve the story. Yet, at the same time, they believed that the simultaneous perception of sound and image in the cinema could ultimately not be transcended and thus had to be acknowledged by the music itself. In this sense, then, sound in the cinema could not claim the same degree of autonomy that other art forms enjoyed, chief among them, autonomous music.[19] Adorno and Eisler thus suggested that the mediated relation of sound and image complicated the question of style. As an analytical category they continued to use the notion of style but they made clear that it had to be adjusted to the medium of the cinema. In Buhler's theoretical overview, however, the question of film style represents the anti-ideological research tradition of neo-formalism. Again, as with the theory of functions, the question of style should not be falsely dismissed as a topic incompatible with critical theories. For obvious reasons, as a well-established approach in musicology, style has been a common concern for the investigation of film music. But shouldn't the notion of style also include other non-musical or non-speech-related forms of "noise" that in recent decades have been getting more and more sophisticated?

Following David Bordwell's approach, style may be defined as "the texture of the film's images and sounds, the result of choices made by the filmmaker(s) in particular historical circumstances."[20] Bordwell goes on to say, "However much the spectator may be engaged by plot or genre, subject matter or thematic implication, the texture of the film experience depends centrally upon the moving images and the sound that accompanies them. The audience gains access to story or theme only through that tissue of sensory materials."[21] Echoing earlier theorists of style, for instance Susan Sontag, Bordwell suggests that every film can be discussed in terms of a "texture," a "tissue of sensory materials" giving the audience access to an experience that is mediated by the texture's components and interplay. As Sontag pointed out in the 1960s, there "is no neutral, absolutely transparent style," thus distancing herself from the traditional divide between content and style in order to make a larger case for art as an expressive form of experience.[22] And as she demonstrated with her discussion of "camp," questions of style could very well be related to questions of ideology—in her case, sexual orientation and subculture.[23] Yet in film studies, even though the neo-formalists have been active proponents of a style-based film history, they have not been much concerned with the history of sound as a distinct aesthetic tradition and have had little to say about sonic styles beyond the various forms of film music. In short, the study of film aesthetics including the history of sound should not be dismissed as anti-ideological. On the contrary, shifting attention from music to all the other sounds, or sounds of the "other," and considering them more systematically as sonic styles in their own right, is a crucial endeavor—to be continued in the next section by returning to the discussion of Robert Altman's films from the early 1970s.

## Robert Altman's *Nashville* and the New Hollywood Cinema

If Steven Spielberg's *Jaws* from 1975 is often seen to have "killed" the New Hollywood Cinema, *Nashville* from the same year may well be regarded as one of its high points. While *Jaws* ushered in the renaissance of the "calculated blockbuster" according to Thomas Schatz, revitalizing the film industry economically, *Nashville* performed a critique of show business and American politics by way of a radically subversive and innovative film aesthetic.[24] Spanning from the late 1960s until the late 1970s, the New Hollywood Cinema thus represents an ambiguous period of innovation and experimentation in American film involving the rebirth of the blockbuster but also an auteur cinema that was inspired by different sources, among them the sophisticated European art cinema of the 1950s and 1960s, the interaction with marginal practices such as exploitation and experimental cinema, and a critical revisionism vis-à-vis classic Hollywood genres shaped by the counterculture.[25] All aspects of film aesthetics, including the soundtrack, were affected by these influences. The influx of pop and rock songs instead of orchestral scores in films such as *The Graduate* (Mike Nichols, 1967) and *Easy Rider* (Dennis Hopper, 1969) was one obvious change, but new sonic styles also emerged in non-musical realms.

As an oppositional aesthetic movement, New Hollywood was feeding on the spirit of the counterculture. However, even though it rebelled against the generic and narrative conventions of mainstream Hollywood, in the realm of sound the norm of synchronization—that is, the understanding that the cinematic apparatus should ensure perfect sync of image and sound—was not its primary target.[26] Its oppositional stance, therefore, did not show itself by using asynchronous sound but rather through unusual or surprising combinations of auditory qualities. To mention just a few examples, in *Easy Rider* the non-linear collage of sound and image during a prolonged drug-trip sequence set in New Orleans combines fragmented voices and machines. *Five Easy Pieces* (Bob Rafelson, 1970) brings together an old piano sound with the honking of cars. *A Clockwork Orange* (Stanley Kubrick, 1971) intensifies a scene of violence through an electronic "making strange" of using the police baton. *The Long Goodbye* (Robert Altman, 1973) includes scenes with the main character mumbling to himself against the background noise. Finally, *The Conversation* (Francis Ford Coppola, 1974) revolves around the obsessive replaying and decoding of a recorded conversation, thus foregrounding the opacity and misleading cues on the soundtrack. All of these examples introduced new and fascinating sonic juxtapositions rather than giving up the norm of synchronization. Prime among these new sound combinations was the clash of the mechanical and the human, the cultured and the ordinary, and the electronic and the natural. In some examples, these new elements of sonic styles only colored individual scenes, while in others they shaped the stylistic texture of the whole film.[27]

This quality of creating not just a few unusual scenes but a new sonic style for a complete film characterizes Robert Altman's work. In the mid-1970s, reviewers noticed that the eight-track system, which Altman had announced so prominently in the credit sequence of *Nashville*, created a fascinating new sonic texture for his film. They called the impression it generated "panoramic" and went on to say: "Altman's sound reaches out across the entire cityscape and records (with non-umbilical radio mikes) a symphony of direct, real, overlapping, synchronized, on-the-spot live sound: a glorious achievement which is guaranteed to thrill filmmakers, technical buffs, and audiences alike, and which has increased the expressive potential of the cinema."[28] For these critics, the sense of innovation and artistic achievement lay not merely in the surprising juxtapositions of sounds but in transforming the separate tracks, which the new technology had produced, into a musical piece, a *symphony* of sound.

From a production point of view, too, multitrack recording signaled a new expressive freedom, and Altman was one of its most ardent practitioners. According to sound recordist Jim Webb, who worked on *Nashville*, the shooting style of filmmaking changed fundamentally because "the improvisation and overlapping of dialogue as it naturally occurs in real conversation whether on or off the screen is accommodated, plus there is the increased capacity to record certain sound effects and sub-conversations that would normally be lost if only the main dialogue were recorded."[29] In addition to producing a larger number of tracks with more varied material, Altman also suggested that multitrack recording partially shifted creative control from the stage of sound recording to the process of mixing: "The eight-track system is simply a way of recording the sound as you shoot it on the set on separate tracks. So it gives you the control to change the balance the way you want it when you finish."[30] The new technology, therefore, allowed Altman to record a multitude of live sounds and voices on the set in separate, unmixed tracks but also to subtly adjust and re-balance them in postproduction, thus enhancing the sense of creative control.

In a short period beginning with *MASH* (1970) and culminating with *Nashville*, Altman used this new artistic freedom to experiment with overlapping dialogue, the inclusion of off-screen conversations, detaching sound from image, and using meta-diegetic sounds.[31] Following Jay Beck's analysis, these experiments contributed to the idea of a "democratic voice" emerging in Altman's films that was characterized by a "non-hierarchical horizontality" and reflected "his method where most, if not all of his characters are given a voice and have the potential to be heard."[32] However, just as the practice of multitrack recording had a downside due to a loss of perspective and mood, the notion of a democratic voice in Altman's films is a complicated aspect of his work.[33] According to film scholar Rick Altman, *Nashville* eschewed Hollywood's tradition of hierarchical sound mixing that historically had been employed to guarantee the transmission of a stable narrative. Instead its multilayered soundscape relied on the contrast between carefully mixed songs and a "cacophonous mode of competing dialogue," giving rise to the feeling of a presence looming in the background of the film as a "mixer-like figure who reigns over image and sound alike."[34] While bringing marginal and off-screen voices into the sound mix was thus more democratic than mainstream Hollywood, the cacophonous mode also signaled a crisis of communication at a more fundamental level, an inability to listen and respond in a genuine exchange. Altman's sonic style of overlapping dialogue may therefore be linked to the metaphor of a democratic voice—establishing an "open narrative structure," as Robert Kolker puts it, or an "interanimation of authentically conflicting voices," according to Helene Keyssar—yet the specifics of this aesthetic need to be discussed at the level of each individual film.[35]

The complex relationship between sonic style and cultural meaning is particularly true for *Nashville*. Combining the setting of Nashville with twenty-four major characters, the performance of music, romance, farce, accidental meetings, and the climax of a political rally including an assassination, the meandering structure of the film aims to capture a cross-section of US-American society. As Patrick McGilligan suggests, "It is not really a plot movie—though it has a plot—and not really a character study. It is more like a 'group portrait' that is also a pseudo-travelogue through a richly detailed slice of Americana."[36] A recurring element in this patchwork of storylines is the sound truck of the presidential candidate Hal Phillip Walker of the Replacement Party whose political announcements are heard throughout the film, emanating in a mechanical and hollow fashion from the loudspeakers attached to the truck. Crisscrossing the streets and speaking, as it were, into the empty spaces of the city, the viewers only get to hear Walker's voice but never to see him in person, not even at the final political rally as the assassination of the singer Barbara Jean (Ronee Blakeley) precludes his arrival on stage. In this way, the sound truck represents a meta-diegetic and acousmatic sound element, which, following Jay Beck, is anchored in the diegetic world but also serves to provide a commentary on the narrated world itself.[37] Giving voice to a grassroots populism dissatisfied with conventional politicians and powerful elites, the need for replacement uttered by Walker's truck in an informal, folksy tone expresses the sense of frustration and anger at the heart of many characters and storylines.

Alongside the musical performances and the characters' often overlapping and cacophonous conversations, this disembodied voice analyzing the political landscape—but speaking into a void and not finding an audience in the film—is another sonic mode employed in *Nashville*'s soundscape. It is subtly related to a visual style characterized by a constantly moving frame and the flat space created by using telephoto lenses. Yet, as becomes clear from the scenes with the truck, the sound does not necessarily follow a logic of realism for which greater visual distance would usually mean a decreasing volume. Rather, the amplified voice of Walker, who does not appear to be physically present in the truck, becomes a pre-recorded monologue floating through the diegesis. In this way, *Nashville* establishes what might be called a shifting or modulated proximity of sounds to visual objects. Human bodies become transitory presences that are sometimes foregrounded aurally and sometimes blocked out or reduced to mute gestures. This is most common in non-musical scenes, but in a sequence at the Nashville Speedway the roaring engines of the race cars drown out a song performed by the character Albuquerque (Barbara Harris), leaving only the visual movements of the guitarist playing his instrument and Albuquerque singing and pantomiming the lyrics of the song.[38]

The practice of unmixing and then rearranging the separate soundtracks that Altman adapted from the music industry, thus, creates a disconcerting and uneasy new balance between visual objects, their sonic qualities and the shifting feeling of closeness. Altman does not abandon the normative practice of synchronization; rather, the eight-track system allows him to remix and rechannel the flow of life into an alternative soundscape. For the viewers, both the modulated proximity to auditory sources and the transitory visual presence of characters make it possible, or rather inevitable, to have only fleeting attachments to the characters and events in the story. Ultimately, the panoramic impression generated by the film allows the audience to get an overview, to constantly move in and out of individual storylines without, however, being able to form lasting bonds. In this sense, the non-hierarchical soundscape of *Nashville* invites an attitude that is at the same time involved and noncommittal, attentive and detached. A world unfolds with constant interaction and noise but without lasting connections or stable relationships.

The sonic and visual style of *Nashville*—a style of modulated proximity and transitory presence—may therefore be connected with the democratic elements of Altman's work in a general sense, but *Nashville* also has more ambiguous implications. In a nutshell, the film's stylistic texture supports the idea of a narcissistic logic of interaction at work in celebrity and everyday cultures, an excessive foregrounding of the self as the ultimate center of attention and interest. Indeed, crucial scenes in the film show the main characters talking to themselves. And yet, it is important to point out that Altman does not explicitly critique this culture of an excessive narcissism. He transforms its underlying dynamic into the stylistic visual and sonic texture of his film, thus forcing the audience to participate in, rather than detach itself from a compulsive form of self-centeredness. In other words, the only way to gain access to the film's plot and themes is through a stylistic texture that offers the logic of narcissism to the viewers in the process of watching the film—creating temporary and fleeting attachments to others but always returning to, and privileging, the individual self.

In this sense, Robert Altman's work, in the larger context of the New Hollywood Cinema, demonstrates that the consideration of sonic properties and functions can and should be related to historical norms at both the level of design and of cultural meanings and ideology. While multitrack technology allowed Altman to orchestrate a multitude of voices in new ways, he chose, in this case, to rearrange them into musical pieces, a cacophony of sounds, and a patchwork of self-centered and often disconnected speech acts, thereby critically but also creatively reconfiguring the representation of conversations in 1970s US-American culture. In more general terms, this chapter has examined, and argued for, the need to historicize the development and transformation of sound in fiction and non-fiction films. It has suggested that the concept of functions may serve as a way of linking questions of film design with the discourse on ideologies in order to integrate auditory styles more explicitly into a history of film style that has been dominated by the visual for too long. If recent publications signal a shift in this direction, much work remains to be done to acknowledge the history of soundscapes beyond the realms of film music or dialogue. And as this discussion of Robert Altman's *Nashville* has suggested, a thorough history of auditory styles will pay close attention to individual sonic properties but also to the texture and the metaphor of mixing as an aspect of film aesthetics as well as cultural ideology.

## Notes

1 Connie Byrne and William O. Lopez, "Nashville [1975]," in *Robert Altman Interviews*, ed. David Sterritt (Jackson: University Press of Mississippi, 2000), 25.
2 Jonathan Sterne, "Sonic Imaginations," in *The Sound Studies Reader*, ed. Jonathan Sterne (London and New York: Routledge, 2012), 1–17.
3 Bill Nichols, "The Voice of Documentary," *Film Quarterly* 36, no. 3 (1983): 17–30. Nichols has continued this discussion, although "voice" often refers metaphorically to the perspective or rhetorical address of the film as a whole rather than the sonic quality of the soundtrack; see Bill Nichols, "To See the World Anew: Revisiting

the Voice of Documentary," in *Speaking Truths with Film: Evidence, Ethics, Politics in Documentary* (Oakland: University of California Press, 2016), 74–89. For a recent collection of essays on sound in documentary films see Geoffrey Cox and John Corner, eds., *Soundings: Documentary Film and the Listening Experience* (Queensgate: University of Huddersfield Press, 2018).
4   For a more elaborate discussion of film music see Christof Decker, "'It flows through me like rain': Minimal Music and Transcendence in *American Beauty* (1999)," in *America and the Musical Unconscious*, ed. Julius Greve and Sascha Pöhlmann (New York and Dresden: Atropos Press, 2015), 187–211.
5   Helen Hanson, *Hollywood Soundscapes: Film Sound Style, Craft and Production in the Classical Era* (London: BFI/Palgrave, 2017), 1.
6   Ibid., 4.
7   See Rick Altman, *Silent Film Sound* (New York and Chichester: Columbia University Press, 2004), Jay Beck, *Designing Sound: Audiovisual Aesthetics in 1970s American Cinema* (New Brunswick: Rutgers University Press, 2016) and Hanson, *Hollywood Soundscapes*.
8   James Buhler and David Neumeyer, *Hearing the Movies: Music and Sound in Film History*, 2nd ed. (New York and Oxford: Oxford University Press, 2016), 12.
9   See David Bordwell and Kristin Thompson, *Film Art: An Introduction*, 9th ed. (New York: McGraw-Hill, 2010), 269–311; Buhler and Neumeyer, *Hearing the Movies*.
10  Mervyn Cooke, "Film Music," in *The New Grove Dictionary of Music and Musicians*, ed. Stanley Sadie, vol. 8 (London: Macmillan, 2001), 806.
11  See Cooke, "Film Music," Mervyn Cooke, *A History of Film Music* (Cambridge: Cambridge University Press, 2008), or Kathryn Kalinak, *Film Music: A Very Short Introduction* (Oxford and New York: Oxford University Press, 2010).
12  James Buhler, "Ontological, Formal, and Critical Theories of Film Music and Sound," in *The Oxford Handbook of Film Music Studies*, ed. David Neumeyer (Oxford and New York: Oxford University Press, 2014), 188–225.
13  Ibid., 207.
14  David Bordwell, "Neo-Structuralist Narratology and the Functions of Filmic Storytelling," in *Narrative across Media: The Languages of Storytelling*, ed. Marie-Laure Ryan (Lincoln: University of Nebraska Press, 2004), 203–19.
15  For a detailed discussion of this model see Christof Decker, "Historicising the Moving Image: Film and the Theory of Cultural Functions," in *Moving Images – Mobile Viewers. 20th Century Visuality*, ed. Renate Brosch (Berlin: LIT, 2011), 75–91.
16  Buhler, "Ontological," 203.
17  Michael Bull and Les Back, "Introduction: Into Sound," in *The Auditory Culture Reader*, ed. Michael Bull and Les Back (Oxford and New York: Berg, 2003), 3.
18  Theodor W. Adorno and Hanns Eisler, *Composing for the Films* (London: The Athlone Press, 1994), 114–33.
19  Ibid., 62–88.
20  David Bordwell, *On the History of Film Style* (Cambridge, MA; and London: Harvard University Press, 1997), 4. However, Hanson's work on Hollywood's sound craft indicates that the notion of "choices made" can become quite complicated. How and why choices were made at the micro level can be difficult, if not impossible, to reconstruct through archival sources, and as the "myriad changes and modifications to sound technologies, style and production practices" at the

synchronic level suggest, trying to determine what constituted the historical norm for specific sound practices can be equally challenging; Hanson, *Hollywood Soundscapes*, 4.
21. Bordwell, *On the History*, 7.
22. Susan Sontag, "On Style," in *Against Interpretation and Other Essays* (New York: Dell, 1969), 25.
23. Susan Sontag, "Notes on 'Camp,'" in *Against Interpretation and Other Essays* (New York: Dell, 1969), 277–93.
24. Thomas Schatz, "The New Hollywood," in *Film Theory Goes to the Movies*, ed. Jim Collins, Hilary Radner, and Ava Preacher Collins (New York and London: Routledge, 1993), 8–36, 265–7; for a different view on the period see Thomas Elsaesser, "American Auteur Cinema: The Last—or First—Picture Show?" in *The Last Great American Picture Show: New Hollywood Cinema in the 1970s*, ed. Thomas Elsaesser, Alexander Horwarth, and Noel King (Amsterdam: Amsterdam University Press, 2004), 37–69.
25. Elsasser, "American Auteur Cinema."
26. See Buhler, "Ontological" for a brief discussion of this norm.
27. For an extensive analysis of the New Hollywood Cinema's soundscape, see Beck, *Designing Sound*.
28. Byrne and Lopez, "Nashville," 21.
29. Jim Webb, with assistance from Don Ketteler, "Using the Multi-Track Format for Production Film Recording," *Recording Engineer/Producer* 11, no. 2 (April 1980): 116.
30. F. Anthony Macklin, "The Artist and the Multitude are Natural Enemies [1976]," in *Robert Altman Interviews*, ed. David Sterritt (Jackson: University Press of Mississippi, 2000), 80.
31. Jay Beck, "The Democratic Voice: Altman's Sound Aesthetics in the 1970s," in *A Companion to Robert Altman*, ed. Adrian Danks, 1st ed. (New York: John Wiley & Sons, Inc., 2015), 426–80, ProQuest Ebook Central.
32. Beck, "The Democratic Voice," 429.
33. According to Jim Webb, the main drawback of multitrack recording was the need to use radio mikes on the individual actors: "This format kills perspective. With multitrack, and its inherent use of radio mikes, each separate dialogue track has excellent quality. [. . .] What is lost is a feeling for the perspective of the environment. And the perspective is the mood whether it be an echoing hallway or a noisy street. It is the thing that gives life to the track. Without it, the movie no longer sounds like it looks." Webb, "Using the Multi-Track Format," 117.
34. Rick Altman, "24-Track Narrative? Robert Altman's *Nashville*," *Cinémas* 1, no. 3 (1991): 110. Jay Beck also mentions that the principle of non-hierarchical horizontality comes at a cost, in particular "the loss of a sense of unity." Beck, "The Democratic Voice," 429, 430.
35. Robert Phillip Kolker, *A Cinema of Loneliness: Penn, Kubrick, Scorsese, Spielberg, Altman* (New York and Oxford: Oxford University Press, 1988), 381; Helene Keyssar, *Robert Altman's America* (New York and Oxford: Oxford University Press, 1991), 5.
36. Patrick McGilligan, *Robert Altman: Jumping Off the Cliff. A Biography of the Great American Director* (New York: St. Martin's Press, 1989), 403.
37. See Beck's analysis of the earlier example *MASH*, "The Democratic Voice," 434–42.

38  Gayle Magee has shown that a number of claims concerning the musical performances (e.g., that the actors had no musical experience) that became pivotal points of marketing the film at the time are not accurate. Many of the main actors had significant experience as musicians or performers and many of the songs existed before the film was shot, ironically linking *Nashville* in this respect to the tradition of classic Hollywood musicals; Gayle Magee, "Songwriting, Advertising, and Mythmaking in the New Hollywood: The Case of *Nashville* (1975)," *Music and the Moving Image* 5, no. 3 (Fall 2012): 28–45.

# Works Cited

Adorno, Theodor W., and Hanns Eisler. *Composing for the Films*. London: The Athlone Press, 1994.
Altman, Rick. "24-Track Narrative? Robert Altman's *Nashville*." *Cinémas* 1, no. 3 (1991): 102–25.
Altman, Rick. *Silent Film Sound*. New York and Chichester: Columbia University Press, 2004.
Altman, Robert, dir. *MASH*. 1970. Twentieth Century Fox. Film.
Altman, Robert, dir. *The Long Goodbye*. 1973. United Artists. Film.
Altman, Robert, dir. *Nashville*. 1975. Paramount Pictures. Film.
Beck, Jay. "The Democratic Voice: Altman's Sound Aesthetics in the 1970s." In *A Companion to Robert Altman*, edited by Adrian Danks, 426–80. 1st ed. New York: John Wiley & Sons, Inc., 2015. ProQuest Ebook Central.
Beck, Jay. *Designing Sound: Audiovisual Aesthetics in 1970s American Cinema*. New Brunswick: Rutgers University Press, 2016.
Bordwell, David. *On the History of Film Style*. Cambridge, MA; and London: Harvard University Press, 1997.
Bordwell, David. "Neo-Structuralist Narratology and the Functions of Filmic Storytelling." In *Narrative across Media: The Languages of Storytelling*, edited by Marie-Laure Ryan, 203–19. Lincoln: University of Nebraska Press, 2004.
Bordwell, David, and Kristin Thompson. *Film Art: An Introduction*. 9th ed. New York: McGraw-Hill, 2010.
Buhler, James. "Ontological, Formal, and Critical Theories of Film Music and Sound." In *The Oxford Handbook of Film Music Studies*, edited by David Neumeyer, 188–225. Oxford and New York: Oxford University Press, 2014.
Buhler, James, and David Neumeyer. *Hearing the Movies: Music and Sound in Film History*. 2nd ed. New York and Oxford: Oxford University Press, 2016.
Bull, Michael, and Les Back. "Introduction: Into Sound." In *The Auditory Culture Reader*, edited by Michael Bull and Les Back, 1–23. Oxford and New York: Berg, 2003.
Byrne, Connie, and William O. Lopez. "Nashville [1975]." In *Robert Altman Interviews*, edited by David Sterritt, 19–33. Jackson: University Press of Mississippi, 2000.
Cooke, Mervyn. "Film music." In *The New Grove Dictionary of Music and Musicians*, edited by Stanley Sadie, 797–810. Vol. 8. London: Macmillan, 2001.
Cooke, Mervyn. *A History of Film Music*. Cambridge: Cambridge University Press, 2008.
Coppola, Francis Ford, dir. *The Conversation*. 1974. Paramount Pictures. Film.
Cox, Geoffrey, and John Corner, ed. *Soundings: Documentary Film and the Listening Experience*. Queensgate: University of Huddersfield Press, 2018.

Decker, Christof. "Historicising the Moving Image: Film and the Theory of Cultural Functions." In *Moving Images – Mobile Viewers: 20th Century Visuality*, edited by Renate Brosch, 75–91. Berlin: LIT, 2011.

Decker, Christof. "'It flows through me like rain': Minimal Music and Transcendence in American Beauty (1999)." In *America and the Musical Unconscious*, edited by Julius Greve and Sascha Pöhlmann, 187–211. New York and Dresden: Atropos Press, 2015.

Elsaesser, Thomas. "American Auteur Cinema: The Last—or First—Picture Show?" In *The Last Great American Picture Show: New Hollywood Cinema in the 1970s*, edited by Thomas Elsaesser, Alexander Horwarth, and Noel King, 37–69. Amsterdam: Amsterdam University Press, 2004.

Hanson, Helen. *Hollywood Soundscapes: Film Sound Style, Craft and Production in the Classical Era*. London: BFI/Palgrave, 2017.

Hopper, Dennis, dir. *Easy Rider*. 1969. Columbia Pictures. Film.

Kalinak, Kathryn. *Film Music: A Very Short Introduction*. Oxford and New York: Oxford University Press, 2010.

Keyssar, Helene. *Robert Altman's America*. New York and Oxford: Oxford University Press, 1991.

Kolker, Robert Phillip. *A Cinema of Loneliness: Penn, Kubrick, Scorsese, Spielberg, Altman*. New York and Oxford: Oxford University Press, 1988.

Kubrick, Stanley, dir. *A Clockwork Orange*. 1971. Warner Bros. Film.

Macklin, F. Anthony. "The Artist and the Multitude are Natural Enemies [1976]." In *Robert Altman Interviews*, edited by David Sterritt, 63–83. Jackson: University Press of Mississippi, 2000.

Magee, Gayle. "Songwriting, Advertising, and Mythmaking in the New Hollywood: The Case of *Nashville* (1975)." *Music and the Moving Image* 5, no. 3 (Fall 2012): 28–45.

McGilligan, Patrick: *Robert Altman. Jumping Off the Cliff: A Biography of the Great American Director*. New York: St. Martin's Press, 1989.

Nichols, Bill. "The Voice of Documentary." *Film Quarterly* 36, no. 3 (1983): 17–30.

Nichols, Bill. "To See the World Anew: Revisiting the Voice of Documentary." In *Speaking Truths with Film: Evidence, Ethics, Politics in Documentary*, 74–89. Oakland: University of California Press, 2016.

Nichols, Mike, dir. *The Graduate*. 1967. Embassy Pictures. Film.

Rafelson, Bob, dir. *Five Easy Pieces*. 1970. Columbia Pictures. Film.

Schatz, Thomas. "The New Hollywood." In *Film Theory Goes to the Movies*, edited by Jim Collins, Hilary Radner, and Ava Preacher Collins, 8–36; 265–7. New York and London: Routledge, 1993.

Sontag, Susan. "Notes on 'Camp'." In *Against Interpretation and Other Essays*, 277–93. New York: Dell, 1969.

Sontag, Susan. "On Style." In *Against Interpretation and Other Essays*, 24–45. New York: Dell, 1969.

Spielberg, Steven, dir. *Jaws*. 1975. Universal Pictures. Film.

Sterne, Jonathan. "Sonic Imaginations." In *The Sound Studies Reader*, edited by Jonathan Sterne, 1–17. London and New York: Routledge, 2012.

Webb, Jim, with assistance from Don Ketteler. "Using the Multi-Track Format for Production Film Recording." *Recording Engineer/Producer* 11, no. 2 (April 1980): 110, 112, 114–17.

Zinnemann, Fred, dir. *High Noon*. 1952. United Artists. Film.

Part II

# Soundtracks of Collective Memory

# 4

# Voice and Wake

## Susan Howe, M. NourbeSe Philip, and the Ecology of Echology

Julius Greve

"Voices I am following lead me to the margins."

—Susan Howe, *The Birth-mark*[1]

"I have often since wondered whether the sounds of those murdered Africans continue to resound and echo underwater. In the bone beds of the sea."

—M. NourbeSe Philip, *Zong!*[2]

What does an echo sound like in the context of contemporary experimental poetry? How should one read it in terms of visual and auditory prosody; in terms of form and content? Do the acoustics of social space evoked by the lyric and the epic alike suggest that form veers toward the visual parameters of a given poem, whereas content is somehow more tied to how it sounds? Does the linkage of sound studies and poetics scholarship necessarily result in the partial reification of auditory prosody by way of connecting it to content, rather than form? And if so, what are the politics connected to such a reification, vis-à-vis the category of voice? Finally, if the inquiry of these issues may be termed *echology*—the discourse on site- and time-specific resounding—in what cultural, historical, and media-technological environment does that inquiry emerge? That is to say: What is the ecology of echology in contemporary poetics?[3]

In the spirit of Nathaniel Mackey's *Discrepant Engagement: Dissonance, Cross-Culturality, and Experimental Writing* (1993), which juxtaposes criticism on African American and Caribbean poetics with essays on the Black Mountain school of American poetry (and, thus, correlates issues of an irreducibly Black poetics with that of postwar experimentalist aesthetics in the United States), I want to explore these issues of echology and/or ecology in the context of two women writers—one Caribbean-Canadian, the other a New England poet. I seek to examine the usefulness of the notion of echology in the work of Susan Howe and M. NourbeSe Philip. To do so, I will closely read Howe's two long poems "Articulation of Sound Forms in Time" and "Thorow" (as published in the 1990 collection *Singularities*) and selected passages from Philip's *She Tries Her Tongue, Her Silence Softly Breaks* (1989) and, in particular, her long poem *Zong!* (2008), which, as its author acknowledges, is partially inspired by the Language poets (meaning, writers such as Charles Bernstein, Bruce Andrews, and Howe herself)—the quintessential representatives of US-American postmodern poetry, in the sense of verse directly inspired by poststructuralist thought.[4]

Given that both Howe and Philip present their poetics with reference to past discourses in and of archival material—of legal and linguistic, philosophical and political, and mythographical and historical sources—I want to compare and contrast their versified takes on specifically American and/or Afro-Diasporic colonial trauma. In light of their highly diverse and cross-genre writings, the homophonic pun of a concept such as echology or, indeed, ec(h)ology, points to the ways in which the trauma of the North American colonial past comes to haunt the present. The genocides of Native Americans and African peoples alike, documented and resounded in works such as "Articulation of Sound Forms in Time" and *Zong!*, respectively, results in the theoretical chiasm of an ineffable and thus unlivable afterlife: What hauntology is to ontology, echology is to ecology.[5] I will explore the ways in which such an afterlife of the colonial period (and also its reflection in nineteenth-century nature writing, such as Henry David Thoreau's *Walden* in the case of Howe's "Thorow")[6] is indexed by what Christina Sharpe calls "being in the wake"—for her an irreducibly Afrocentric, "Black" condition of possibility (or, rather, impossibility) with respect to discourses of oppression. As she puts it, this condition describes, among other things, the awareness of "antiblackness" as "the ground on which we stand, the ground from which we attempt to speak" (while one of the definitions she gives for the notion of "the wake" itself is *"the track left on the water's surface by a ship; the disturbance caused by a body swimming or moved, in water"*: a definition that will become meaningful in the discussion of Philip's poetry in part 2 of this chapter).[7]

Yet, in the larger context of a poetics of the margin or the minor, and again in the spirit of Mackey's cross-cultural approach to a contemporary *poiesis* of "dissidence and experimentation," of "writers who, poet or novelist, black or white, from the United States or from the Caribbean, produce work of a refractory, oppositional sort,"[8] this approach may prove fruitful with regard to both poets at hand, not just in the case of Philip. This hypothesis is based on the generic history of the lament and its voicing in contemporary experimental writing (very much including the work of Howe and Philip), and it is based on the echopoetic condition of marginal writing qua being a revision of or response to an oppressive antecedent: that is to say, modernism and modernity. Voice and wake, echo and *oikos*: both Howe and Philip evoke a past that is unbearable, a present that must or ought to "defend the dead."[9]

## Regionalist Acoustics

"Articulation of Sound Forms in Time," the first section of poems in Howe's *Singularities*, chronicles the wanderings of Reverend Hope Atherton after a massacre of Native American men, women, and children in the year 1676. Directly after the ending of "King Philip's War so-called by the English," Native Americans, including "Squakeags, Pockomtucks, Mahicans, Nipmunks, and others,"[10] had made camp near Deerfield, in the Connecticut River valley, and allegedly threatened the settlers in that area. The consequence of this gathering was the colonial force sending troops of roughly 160 men, with whom Reverend Atherton went, in order to defend the settlers who were supposedly in danger. The troops, during their march from Hatfield, had encountered a small, unprotected camp, which was subsequently destroyed, with the women, children, and men having been brutally massacred. After the regrouping of Native American warriors and the reclaiming of their own space, Atherton flees into woods with a few other survivors, eventually finding his way back home. His contemporaries do not take seriously the stories about his experiences with the Natives during his errantry, which he has put into writing. He is deemed a madman, due to his harrowing account of the Native American warriors burning alive the lost survivors from the previous battle. As Howe surmises in the first part of her poem, called "The Falls Fight," which serves as an introductory prose piece to the subsequent verse, this is "Hope's baptism of fire. No one believed the Minister's letter. He became a stranger to his community and died soon after . . ." in obscurity.[11]

In an early essay on Howe's lyrical reframing of the massacre in the unguarded Native American camp in 1676, Rachel Blau DuPlessis contends that it "can be read as an allegory of how the center, how major man—white, colonist, Protestant, male . . . —how that man, entering almost accidentally some marginal space goes from the straight and narrow to sheer errancy, sheer wanderings."[12] The allegorical dimension of Howe's poetic practice in "Articulation" is also connected to the etymology of the name "Hope" itself: "In our culture Hope is a name we give women," and "Pre-revolution Americans viewed America as the land of Hope." And finally: "I assume Hope Atherton's excursions for an emblem foreshadowing a Poet's abolished limitations on our demythologized fantasy of Manifest Destiny."[13] The speaker (given that in Howe's multi-genre approach to writing, poetic form includes both verse and prose) as well as the capital-P "Poet" herself establishes the predetermined violence done to Native Americans, to the ideal of Puritan religious dissidence, as well as by *and to* the archive in the form of this postmodern "Articulation of Sound Forms in Time." For Howe, the—importantly female—Poet must disassemble the discursive, cultural, and historical structures that have led to multiple forms of violence; resulting, as it does, in the resounding, the echoing of the reverend's account and also the discourses that have enveloped it from the beginning. Furthermore, what Gerald Bruns has said of Howe's later work equally applies to "Articulation": "Howe's work is a project of self-formation through the appropriation of the writing (and therefore the subjectivity) of others."[14] How does such a reappropriation of structural violence via the subjectivities of historical figures work on the level of auditory prosody?

The allegorical function of "Hope" is to bemoan that which is lost in the course of the colonial enterprise on North American soil, via a general "baptism of fire."[15] Yet this proper name is also the "emblem" of the postmodern woman writer's stance of having "abolished limitations" from the start. In accord with its definition of poetic form for a title,[16] the first section of Howe's text is to be read as a declaration of ambivalence, insofar as it objects to the originary violence of the Puritan and colonial past of the US, *and* embraces the place in which that violence originated, in the form of her regionalist acoustics of the archive (meaning, her carefully crafted archive-based verse, which not only seeks to reproduce sections of its sources in visually intriguing ways, but also puts an emphasis on the auditory aspect of these sections, in the context of poetic form). Arguably, it is this ambivalence that is at the heart of Howe's poetic practice *qua* echology (of her resounding of regional histor(iograph)y in verse), an ambivalence that is captured most concisely in the oft-cited line from the poem: "Collision or collusion with history."[17] Consider Marjorie Perloff's gloss of that line: "What a difference a phoneme makes! One's *collision* with history may be accidental, an encounter of opposed ideas neither planned nor anticipated. One's *collusion*, on the other hand, is by definition pre-meditated." She goes on by arguing that "[a]ttentiveness to such difference *(/i/* versus */uw/)* has always distinguished Susan Howe's 'history poems' from those of her contemporaries."[18] While I do not want to argue for or against the higher level of attentiveness to linguistic detail in Howe as opposed to Lyn Hejinian or Bernstein, I suggest that there is an allegorical dimension at work that has "Hope" mirror "Howe" in terms of the heritage of colonial violence done to Native American women, men, and children, and the larger intellectual frameworks of Enlightenment thought that, as "Articulation" suggests, are part and parcel of that heritage. The complicity in that violence, in the final analysis, is reworked on the page of the poem in the form of its own drifting from one discursive field to the next. I claim that this allegorical dimension is to be regarded in parallel with the juxtaposition of "collision" and "collusion": Reading the slanted parallel between "Hope/Howe" and "collision/collusion" as the conceptual scaffolding of "Articulation"—one being a near homograph, the other a near homophone—means taking seriously the echo of the first term in the second. Such a reading, grounded in the regionalist echology of Howe, is confirmed, in particular, when paying close attention to the line preceding the one Perloff singles out, with which the latter forms a couplet: "Predominance pitched across history / Collision or collusion with history."[19] It is also confirmed when considering other couplets that work in a similar chiasmic fashion: "Shape of some many comfortless / And deep so deep as my narrative," from the second section of the poem, "Hope Atherton's Wanderings," or, elsewhere, "Colonnades of rigorous Americanism / portents of lonely destructivism / Knowledge narrowly fixed knowledge / Whose bounds in theories slay," from the third and final section, titled "Taking the Forest."[20] Irrespective of the graphically sophisticated forms of Howe's palimpsests, singled out by a host of scholars and based, to be sure, on the fact of her being a painter by profession before transitioning into writing poetry in the 1970s, the aspect of sounding or resounding, of echoing differently that which is "pitched across history," is at the heart of her archival acoustics.[21]

Like the first long poem of *Singularities*, "Thorow" too opens with a prose section, however unnamed. The title itself, unmistakably alluding to the nineteenth-century transcendentalist Thoreau, also implies a slanted, or an altered, pronunciation of the word "sorrow," that which is expressed in genres such as the lament or the elegy.[22] Sorrow felt for what or whom; by whom? In the poem's introductory section, Howe (or the speaker) relates how in early 1987 she was a writer-in-residence at Lake George in upstate New York, living in a cabin near the lake. She abhors the "travesty . . . of a town," a sad display of the process of modernization and postmodernization, with "the inevitable McDonald's, a Howard Johnson, assorted discount leather outlets, video arcades, a miniature golf course . . . a Donut-land, and a four-star Ramada Inn built over an ancient burial ground."[23] She then asks rhetorically: "what is left when spirits have fled from holy places? In winter the Simulacrum is closed for the season."[24] Yet she does not keep on reminding her reader of the Baudrillardian demise of American culture in the late 1980s. In a subsection called "Narrative in Non-Narrative" she describes how, in the seventeenth century, the area of the lake was discovered and then colonized by "[p]athfinding believers in God and grammar" who "spelled the lake into *place*. . . . In paternal colonial systems a positivist efficiency appropriates primal indeterminacy."[25] Quickly it becomes clear that this ethos of an ineffable non-reified reality, both pre- and post-colonization through armed force and systematizing grammar—through imperialism and Enlightenment, that is—is what is at stake in section 2 of *Singularities*. It has to be noted, too, that the century referred to is the same as that of the massacre of Native Americans and the errantry of Atherton in "Articulation." Both long poems explore nearly congruent themes: "history, system, and authority," and "the mutual embeddedness of colonial power and language," as Will Montgomery puts it.[26] Like the versification of Atherton's sojourn, and like its reflection on the reciprocal relationship of religious and colonialist enterprise, "[t]he wanderings of the poet and speaker in this poem are complicit with *and* in opposition to the European exploration and settlement of this place," as Jenny White states.[27] Thus, a re-casting of the place as unnamed is at stake, precisely by echoing its equally violent and graceful history; violent in terms of "positivist efficiency" and graceful in terms of "primal indeterminacy."[28]

And it is Gilles Deleuze and Félix Guattari who are summoned by Howe to replace Jean Baudrillard, in search of that emplaced grace: she quotes the following lengthy passage from *A Thousand Plateaus* (1980):

> The proper name (*nom propre*) does not designate an individual: it is on the contrary when the individual opens up to the multiplicities pervading him or her, at the outcome of the most severe operation of depersonalization, that he or she acquires his or her proper name. The proper name is the instantaneous apprehension of a multiplicity. The proper name is the subject of a pure infinitive comprehended as such in a field of intensity.[29]

What is the use of this passage in the context of "Thorow"? What is its echological function; its role within Howe's regionalist acoustics? While the authors of this passage share the disdain for certain schools of philosophy grounded in systematization and taxonomy, what is more striking is how the poet goes on by quoting from Thoreau's reflections on Native American names for their environment's places, "rivers, lakes, &c.," and ends her introduction to "Thorow" by proposing a poetological statement similar to those from the beginning of "Articulation," this time steeped in the concepts of Deleuze and Guattari: "Every name driven will be as another rivet in the machine of a universe flux"—and in truly Steinian fashion ending the paragraph on an absent period, thus suggesting a meeting of the content (the activity of "a universe flux") with its form (a sentence without end or, rather, end point).[30]

While the next page features more quotes by Thoreau on Native American names and by contemporaries of his, the main body of the poem paraphrases sections of *Walden* and reflects the discourse and history of domination mentioned above, vis-à-vis the New England ecology of Howe's, Thoreau's, and the seventeenth-century pathfinders' era, resulting in couplets and single lines such as these:

> Thaw has washed away snow
> Covering the old ice
> . . .
> Armageddon at Fort William Henry
> Sunset at Independence Point
> . . .
> Maps give us some idea
> Apprehension as representation
> . . .
> I have imagined a center
>
> Wilder than this region
> The figment of a book[31]

Or else, the echo of (in the sense of "auditory variation on") that theme of mapmaking, of demarcating:

> The expanse of unconcealment
> So different from all maps
> . . .
> Dark here in the driftings
> In the spaces of drifting
>
> Complicity battling redemption[32]

In what sense is this echology a form of lamentation? How can we think through its poetological parameters as meaningful even beyond the level of intertextuality? And how does this form of lyricism qualify as "being in the wake"?[33] Both "Articulation" and "Thorow" lament the turn of US-American beginnings qua violent conquest of ecological and social relations. It is a form of poetic practice that is more than a textual mirroring of sources—despite its postmodern cultural context of fellow writers and artists, and despite Howe's penchant for archival research—precisely because of its detail on the level of auditory prosody, noted by Perloff early on.

Montgomery, too, argues along those lines, proposing a slightly different conception of textual (and extratextual) influence in regard to "Thorow" and its (equivocal) bemoaning the loss of the New England wilderness in a couplet from the second part of the poem: " 'The literature of savagism/ under a spell of savagism' (*Thorow*, 49) could be read as an acknowledgment of Thoreau's theory of the 'wild savage' within and its relation to the wildness in literature. . . . However, read in the context of Thoreau, it becomes clear that Howe's return to origins is only masquerading as such; it is better understood as an act of literary ventriloquism,"[34] that is, a dismissal of the poetic act that would work as an ethical statement. Howe's literary ventriloquism, rather, means the echoing (or partial embodiment) of past collective trauma and its voices via one's own lived experience—as a writer-in-residence, as a woman who is a Poet; as a writer in the wake of a patriarchal or paternalist canon of postwar American poetry.[35]

"There is no recourse to the possibility of a prelapsarian language," Montgomery writes, "yet poetry, in evading the instrumentalized language of commercial exploitation, contains the potential to voice the uncolonized areas of the psyche that Thoreau considers the domain of the savage."[36] Recalling Deleuze and Guattari, it is possible to argue that Howe, in her re-embodiment and simultaneous displacement of New England's past voices and forms of violence, constructs her proper name as poet by means of depersonalization: In this sense ventriloquism as a poetic practice is also a disembodiment of the colonial past, as it is a replacement of its voices—the poetic act itself including the possibility of an ethics, but not presuming the latter as condition of possibility.

## Echological Environments

Christina Sharpe's book *In the Wake: On Blackness and Being* (2016) contains a discussion of M. NourbeSe Philip's *Zong!* that aims to surmise the ethos of this long poem: "What does it look like, entail, and mean to attend to, care for comfort, and defend, those already dead, those dying, and those living lives consigned to the possibility of always-imminent death, life lived in the presence of death; to live this imminence and immanence as and in the 'wake'?"[37] To be sure, compared to the poetry contained in Howe's *Singularities* this concept of "being in the wake," or rather, of "Black being in the wake,"[38] has a different, and more direct resonance with the work of Philip; not just because of its direct examination in Sharpe's monograph. Nonetheless, Howe's echology understood as a kind of literary ventriloquism, to use Montgomery's phrasing again, is certainly related to the structurally and literally violent backdrop of colonial, and postcolonial (in the sense of post–Civil War) North American culture and politics that is at stake in Philip's 2008 long poem.

Already in the 1989 book *She Tries Her Tongue, Her Silence Softly Breaks* (published in the same year as Howe's long poem on Atherton's errantry), Philip concerns herself with the reciprocity of linguistic and cultural appropriation in the wake of centuries of Black enslavement. Consider, for instance, these opening lines of the tellingly named poem "Discourse on the Logic of Language":

> English
> is my mother tongue.
> A mother tongue is not
> not a foreign lan lan lang
> language
> l/anguish
>     anguish
> —a foreign anguish.
>
> English is
> My father tongue.
> A father tongue is
> a foreign language
> . . .
> not a mother tongue.
> . . .
> I have no mother
> tongue[39]

While these lines certainly negotiated the abovementioned reciprocity, it is also the pained repetition of the first syllable of the word "language" *in that very same language*—English—that reflects one of the ways in which one might speak of an echology in Philip's poetics: the resounding—in other words, the repetition—of historical trauma on the level of language *by means of the latter* (thus, not merely implying a reciprocal determination of linguistic and cultural appropriation on the level of content, but a recursive approach to the critique of language on the level of form—that is to say, *sound*—since, after all, the critique is aimed at its object by means of that object). This might be one reason why George Elliott Clarke, in reviewing Carol Morrell's 1994 anthology of Afrodiasporic Canadian women's writing, *Grammar of Dissent: Poetry and Prose by Claire Harris, M. NourbeSe Philip, Dionne Brand*, writes of what he calls "political ventriloquism" that "allows the writers 'both a community and a coherent sense of self—however fictive or imaginative—from which to act and write.'"[40] This type of ventriloquism—literary and political in equal measure—has the speaker assuming the grief and loss of multiple generations of slaves and displaced people of color, whose experiential trajectories have been dominated by the colonial and Enlightenment projects of the same patriarchal order Howe's voicings in *Singularities* recount and cast anew, differently. And while Clarke's phrase aptly points to Philip's political disembodiment through verse and contextualizes her own acoustics of social *poiesis* in her early work, it is especially in terms of her highly acclaimed long poem *Zong!* that the echological metaphor of ventriloquism as a poetic practice becomes meaningful.

At this point I want to rephrase the set of issues listed in the beginning of this chapter; namely, those dealing with sound, prosody, reification, experimentalism, and, finally, echology: What does the duality of visual and auditory prosody amount to if the poem at hand involves historical phenomena such as the Middle Passage, the multiple deaths of enslaved people, and the entanglement of several environments marred by trauma (the slave ship, the sea, the court)? How to resound the traumatic tensions on a social, economic, or ecological level, if their representation seems out of the question, an impossible task? Philip's *Zong!*, which the poet herself has described as "the Song of the untold story," which "cannot be told yet must be told, but only through its un-telling,"[41] reframes the historical trajectory of the British slave ship *Zong* whose captain and crew in late 1781 (and thus roughly a century after the events depicted in Howe's regionalist long poems examined earlier) let drown approximately 150 African slaves on their way to the Caribbean.[42] With the owners of the slave ship trying to collect money from their insurers and being rejected by the latter, legal battles over the incident of the *Zong* resulted in the *Gregson v Gilbert* (1783) case, the account of which Philip (who herself is a lawyer by profession) uses as the primary source material for her poem. In her "un-telling" of a legal document in poetic form, *Zong!*'s author reenvisions and resounds the grief of generations over the court's decision that the killing of human beings may be regarded as "a throwing overboard of goods."[43]

Similar to the case of Howe, I argue that it is key to examine Philip's literary demarcation of bodies, materials, and texts related to the *Zong* incident with an emphasis on the work's auditory prosody, rather than its visual intricacy. Philip's poetics does not only critically evoke the horrors of the transatlantic slave trade on the level of page space, but it also generates what could be called *echo*logical environments in which the discourse of legality rings hollow. In this sense, Philip wards off the question of representation by ventriloquizing the very document that purports to determine the nature of human beings as "goods."

How can we determine an echological environment in verse and what are the ethical stakes of such a literary and political ventriloquism? As in her earlier poems, Philip works with the stretching of words, the repetition of letter or syllables to aim at an auditory effect of stammering or echoing in the environment in which the specific communicative situation of the poem is set: in the case of "Zong! #1," the very first poem of section one, which is called "Os" (meaning, *bone* in Latin),[44] we get a poetic structure such as the one displayed in this chapter's appendix.

Such a reading track, lets the reader both search and combine the individual words, and lines of verse, and suggests, almost like a would-be transcript, the unimaginable sounds made by the individuals thrown overboard by the slaver. This poetic form recalls the ethical investment, which is, according to Sharpe, at the heart of Philip's endeavor, namely: "What does it look like, entail, and mean to attend to, care for comfort, and defend, those already dead, those dying, and those living lives . . . in the presence of death; to live this imminence and immanence as and in the 'wake'?"[45] The ethical thrust of Philip's echology, pointing out slavery's afterlives in the historical moment of *Zong!*'s publication by a Caribbean-Canadian Afrodiasporic author in the first decade of the twenty-first century, is confirmed by that author herself, who writes that in her poem, those whose voices muted by oceanic depths become endowed with a voice by means of a recasting of the verdict that deems them goods: "In *Zong!*, the African, transformed into a thing by the law, is re-transformed, miraculously, back into human. Through oath, and through moan, through mutter, chant and babble."[46] The ethical thrust of displaying the word "water" in the fashion displayed in the appendix is meant to be a re-humanization of that which, by law, had been made a commodity, a "good" to be traded. And yet, similar to Howe's complicity or, rather "collusion" with the sources she recasts in her work, what—in the spirit of the author of *Specters of Marx*—could be named etho-echological writing, regarding the stance adopted by the poet, becomes a vexed issue. For Veronica Austen, the central inquiries of the poem are, first of all, whether its author may "ethically assume the right to tell the story of the *Zong*"[47]—an issue that can also, or even more so, be asked in the cases of "Articulation of Sound Forms in Time" and "Thorow." Furthermore, "how can readers be positioned to become secondary witnesses to the event, and thereby experience a heightening of their social conscience, yet all the while be positioned to respect the ultimate unknowability of the event?"[48] In other words, how could it be possible to adequately represent a kind of cultural and historical trauma, the particularity of which is ineffable, yet whose generality is still at work centuries on, precisely "in the wake" of that trauma, or in "the afterlife of slavery"?[49]

Philip's increasingly scattered writing style, which seems at times to depict the slaves' bodies in the belly of the ship, and which, from the section called "Ebora" ("underwater spirits" in Yoruba) onwards,[50] is characterized by grey and often unreadable typeset, both transforms, and is constrained by, the very language used in *Zong!* The author concedes: "In the discomfort and the disturbance of the poetic text, . . . I implicate myself. The risk of contamination" or, in Howe's terms, of complicity, "lies in piecing together the story that cannot be told."[51] Surely, such an admission would complicate the ethos singled out by Sharpe in her reading of the poem. And yet, it is in the auditory prosody that the ethical thrust of the traumatic environment of the slaver *Zong* via its literary echology in the form of *Zong!* comes to the fore: "Philip's style may offer the possibility of linear progress," Austen reasons; that is, a conventional reading movement from "left to right, top to bottom," and so on; and yet "the surfacing of alternate words and reading paths constructs a text with multiple layers of significance. Furthermore, to hear Philip orally perform the text often reveals a preservation of the fragmentation; in Philip's style of reading, [a word such as] 'the y,' for instance, would not necessarily be 'they,' but 'the why.'"[52] And this is why, I argue, the ethical dimension of Philip's endeavor consists, above all, in the auditory, the performative aspect of the poem (which, as per Howe's literary ventriloquism, is different from saying that the poetic act would be identical to, or based on an ethical statement): this much, after all, is suggested in the "Ebora"-section, one of whose chief phrases initiating this most unreadable, and hence practically unknowable (on the level of page space, that is), concluding part of this experiment in poetic rehumanization of those in chains:

... this is but an oration   of loss ... .[53]

## The Disembodiment of Lived Experience

If Howe's project is to be seen as a depersonalization of the Poet in search for her proper name in the wake of a colonial past that mutes its margins, Philip de-commodifies the human beings divested of their status as women, children, and men by that same language that is used in law and thus in the poem. It is only fitting, then, that both Howe, whose archival "Voices I am following lead me to the margins,"[54] and Philip, whose "*Zong!* is the Song of the untold story," which "cannot be told yet must be told, but only through its un-telling,"[55] have repeatedly performed—that is, sung, chanted, spoken—the poems examined in the present chapter.[56] Indeed most recently, Howe has, together with artist Shannon Ebner and poet Nathaniel Mackey—the author of the aforementioned *Discrepant Engagement*—released the LP *STRAY: A GRAPHIC TONE*, which features recordings of crucial parts from "Articulation" and "Thorow," clearly signaling the importance of those pieces within the poet's oeuvre.[57] In opening the gatefold of the vinyl case, some of the poems by Howe and Mackey are reproduced. Among the more recent poems recorded for *STRAY* is the Bing Crosby tune "Sir Little Echo," the lyrics of which are featured in Howe's 2017 book of poems, *Debths*.[58] Above all, however, the gatefold contains an interview excerpt with Howe in which she states that Mackey's and her "work shares a sense of dispersion in space. I mean, because ultimately ... *sound* is everything in poetry. It's measure; the measure is everything, even though I have gone on about silence ... Every mark on paper is an acoustic mark. Sound is also, obviously, sight. It's that instant flash of recognition that echoes and reechoes."[59]

At least one of the central questions raised in the beginning of this chapter—does the linkage of sound studies and poetics scholarship necessarily result in the reification of auditory prosody by way of connecting it to content, rather than form?—can be restated and provisionally answered at this point: What is the function of the apparent prevalence of auditory prosody in the reading and analysis of visually intricate forms of contemporary experimental poetry? The function of underlining such a prevalence is to take seriously the ethico-aesthetic force of voicing "sound forms in time" and of conveying the sensation of "being in the wake," however unbearable. Along these lines, an always already time- and site-specific (and at times even regionalist) poetics can be witnessed in the work of both Howe and Philip. Both poets work according to an aspiration to embody and distort the voices of past collective trauma, and, therefore, to re-place and disentangle their sounds across time and history, across culturally specific spaces and historiographies. Ultimately, while a general definition of an echological environment would belie its context-specific disposition, in the writings of Howe and Philip, the ecology of echology certainly entails the disembodiment of lived experience by means of experiments in verse.

# Appendix

*Zong! #1*

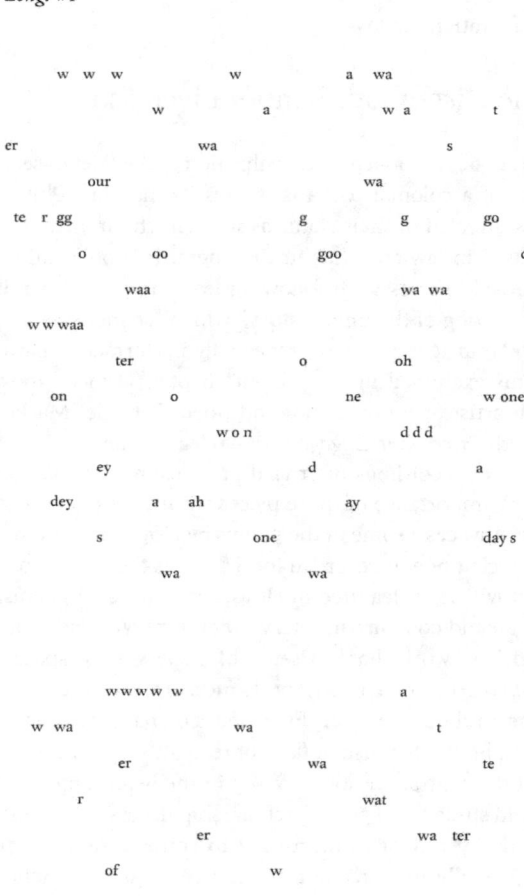

**Appendix** Credit: Philip, M. NourbeSe. Excerpts of *Zong!* © 2008 by M. NourbeSe Philip. Published by Wesleyan University Press and reprinted with permission.

# Notes

1. Susan Howe, *The Birth-mark: Unsettling the Wilderness in American Literary History* (New York: New Directions, 1993), 4.
2. M. NourbeSe Philip, *Zong!* (Middletown, CT: Wesleyan University Press, 2008), 203.
3. In many ways, this account of "echology" is congruent with what Beth Ann Staley has recently conceptualized in her work on Emily Dickinson. In fact, she has

already explored the duality of resonance and environment in the nineteenth-century poet: "Neither a trope nor a technique, Dickinson's art as echo, what I call 'echology' combines the terms 'echo' and 'ecology' to study the ecology of the echo that, when it's assigned to perception instead of narrative, exposes the continuum of the senses and tells us something about the condition of that continuum too" (Beth Ann Staley, "Emily Dickinson's Echology: A Listener's Reconceptualization of Citizenship, Consciousness, and the World" [PhD-Thesis, West Virginia University, 2019], 29). Nonetheless, it has to be noted that Staley neither points out the importance of Howe's reading of Dickinson in *My Emily Dickinson*, nor does she examine twentieth-century poets more generally with respect to the echological aspect of ecopoetics; nor, for that matter, does she explore such an aspect regarding contemporary forms of poetic ventriloquism, as I will do vis-à-vis Howe and Philip.

4   See Philip, *Zong!*, 197. Consider, in this context, Bernstein's own concept of "echopoetics," which he defines as "the nonlinear resonance of one motif bouncing off another within an aesthetics of constellation. Even more, it's the sensation of allusion in the absence of allusion . . . a shadow of an absent source . . . Echopoetics is dialogic critical encounters" (Bernstein, *Pitch of Poetry* [Chicago: University of Chicago Press, 2016], x). In any case, this Bernsteinian notion points to the critical practices of both Walter Benjamin ("constellation") and Mikhail Bakhtin ("dialogic"). Pragmatically speaking, in his book *Pitch of Poetry*, "Echopoetics" merely is the title of selected interviews. In conceptual terms, "echopoetics" is an apt notion for the postmodern lyrical practice of that particular writer associated with Language poetry who has consistently worked with archival materials; meaning, Bernstein's concept is certainly an important inspiration for my present discussion of Howe's work. See, on the uneasy relationship between Language poetry and Howe's poetics, Mandy Bloomfield, *Archaeopoetics: Work, Image, History* (Tuscaloosa: The University of Alabama Press, 2016), 42.

5   Jacques Derrida's coinage of "hauntology" is apt in this equation insofar as Philip repeatedly cites *Specters of Marx* in "Notanda" (her postscript to *Zong!*), the book which elaborated on the downfall of communism and, thus, the state of Marxism: "this element itself is neither living nor dead, present nor absent: it spectralizes. It does not belong to ontology, to the discourse on the Being of beings, or to the essence of life or death. It requires, then, what we call . . . *hauntology*" (Derrida, *Specters of Marx: The State of the Debt, the Work of Mourning and the New International* [New York: Routledge, 2006], 63). This is Philip's explicit admission of Derrida's influence on (her thinking about) *Zong!*: "Re-reading *Specters of Marx* by Derrida has clarified some of my own thought and confirmed me in my earlier feelings that *Zong!* is a wake. It *is* a work that employs memory in the service of mourning—an act that could not be done before . . . Our entrance to the past is through memory. And water. It is happening always—repeating always, the repetition becoming a haunting . . . I come—albeit slowly—to the understanding that *Zong!* is hauntological" (Philip, *Zong!*, 202, 203). In the context of these references, the opposition of "auditory culture" and "sound studies"—that is to say, discourse-analytical forms of musicology vs. the ontologically informed accounts of sound cultures—seems to be a false dilemma: see Brian Kane, "Sound Studies without Auditory Culture: A Critique of the Ontological Turn," *Sound Studies* 1, no. 1 (2016): 2–21.

6. Regarding the intertextual connection between *Walden* and Howe, see: Jenny L. White, "'The Landscapes of Susan Howe's 'Thorow,'" *Contemporary Literature* 47, no. 2 (2006): 245.
7. Christina Sharpe, *In the Wake: On Blackness and Being* (Durham, NC: Duke University Press, 2016), 45, 7, 3.
8. Nathaniel Mackey, *Discrepant Engagement: Dissonance, Cross-Culturality and Experimental Writing* (Cambridge: Cambridge University Press, 2009), 1.
9. Philip, *Zong!*, 25.
10. Susan Howe, *Singularities* (Middletown, CT: Wesleyan University Press, 1990), 3.
11. Ibid., 4.
12. Rachel Blau DuPlessis, "'Whowe': An Essay on Work by Susan Howe," *Sulfur* 20, no. 1 (1987): 163.
13. Howe, *Singularities*, 3.
14. Gerald L. Bruns, "Voices of Construction: On Susan Howe's Poetry and Poetics," *Contemporary Literature* 50, no. 1 (2009): 28.
15. Howe, *Singularities*, 3.
16. See Howe's comments about her understanding of poetry in a 1989 interview, reprinted in *The Birth-mark*: "Poetry is a sort of music. And then I think that the first experience we probably have of the world . . . is sound. We are slapped and we cry. Before we know what meaning is. So to be born would be to hear sound you couldn't understand. And to die is to hear sound, then silence. So it's the articulation that represents life. And Hope has that sort of experience. And Hope is in me. In all of us" (Howe, *The Birth-mark*, 172). It is easy to see how the themes of/in "Articulation" function both on a cultural-historical level and on that of lived experience. This is also how I would read the ambivalence of the word "Forms" in "Sound Forms in Time"—it has to remain up for discussion whether it is meant to be a noun or a verb.
17. Howe, *Singularities*, 33.
18. Marjorie Perloff, "'Collision or Collusion with History': The Narrative Lyric of Susan Howe," *Contemporary Literature* 30, no. 4 (1989): 518.
19. Howe, *Singularities*, 33.
20. Ibid., 9, 12.
21. See, for instance, Perloff, " 'Collision or Collusion with History': The Narrative Lyric of Susan Howe," 519.
22. See, on this matter Eleni Ikoniadou's gloss of the genre: "All over the ancient and modern world, death is a woman's business. Women wash, dress and decorate the corpse, and then sing it to its final resting place with a lament. Lamentation is an extreme expression of sorrow that precedes every other form of oral ritual, and has led to the creation of the oldest epic poems across human culture" (Ikoniadou, "The Lament," in *Unsound: Undead*, ed. Steve Goodman, Toby Heys, and Eleni Ikoniadou [Falmouth: Urbanomic, 2019], 73). I read the second part of Howe's *Singularities* precisely in this manner, while it is at the same time a poetic instance of depersonalization, as my analysis will show. (Note, in this regard, that an additional definition of "the wake," as Sharpe shows, is the following: "*Wake: a watch or vigil held beside the body of someone who has died, sometimes accompanied by ritual observances including eating and drinking*" [Sharpe, *In the Wake: On Blackness and Being*, 10].)
23. Howe, *Singularities*, 40.

24  Ibid.
25  Ibid.
26  Will Montgomery, *The Poetry of Susan Howe: History, Theology, Authority* (New York: Palgrave Macmillan, 2010), 95, 98.
27  White, "The Landscapes of Susan Howe's 'Thorow,'" 257; emphasis added.
28  Howe, *Singularities*, 40.
29  Gilles Deleuze and Félix Guattari, *A Thousand Plateaus: Capitalism and Schizophrenia*, trans. Brian Massumi (London: Continuum, [1980] 2004), 42.
30  Howe, *Singularities*, 41.
31  Ibid., 51, 54.
32  Ibid., 55. See, again, White, "The Landscapes of Susan Howe's 'Thorow,'" 257, on this passage.
33  Sharpe, *In the Wake: On Blackness and Being*, 45.
34  Montgomery, *The Poetry of Susan Howe*, 105.
35  Note, in this regard, the poet's statement, "I am suspicious of the idea of a canon in the first place because to enter this canon a violation has usually been done to your work, no matter what your gender may be" (Howe, *The Birth-mark*, 170). This statement both signals the frustration regarding the patriarchal dynamics of literary and cultural production, and it troubles the straightforward ascription of Howe's work to feminist strands of thought. On the latter issue, see: Ming-Qian Ma, "Articulating the Inarticulate: *Singularities* and the Counter-Method in Susan Howe," *Contemporary Literature* 36, no. 3 (1995): 470.
36  Montgomery, *The Poetry of Susan Howe*, 105.
37  Sharpe, *In the Wake: On Blackness and Being*, 38.
38  Ibid., 45.
39  M. NourbeSe Philip, *She Tries Her Tongue, Her Silence Softly Breaks* (Middletown, CT: Wesleyan University Press, 2014), 30.
40  George Elliott Clarke, *Odysseus Home: Mapping African-Canadian Literature* (Toronto: University of Toronto Press, 2002), 259; Carol Morrell, "Introduction," in *Grammar of Dissent: Poetry and Prose of Claire Harris, M. NourbeSe Philip and Dionne Brand*, ed. Carol Morrell (Fredericton: Goose Lane Editions, 1995), 10.
41  Philip, *Zong!*, 207.
42  Veronica Austen states that it was "[a]pproximately 132 in total," yet, in following Philip's estimation, concedes that the exact number is unclear. "With reportedly some, but obviously not enough, objection by the crew, this mass murder took place over three days" (Veronica J. Austen, "*Zong!*'s 'Should we?': Questioning the Ethical Representation of Trauma," *ESC* 37, nos. 3–4 [2011]: 62).
43  Philip, *Zong!*, 210.
44  Ibid., 3.
45  Sharpe, *In the Wake: On Blackness and Being*, 38.
46  Philip, *Zong!*, 196.
47  Austen, "*Zong!*'s 'Should we?': Questioning the Ethical Representation of Trauma," 64.
48  Ibid. Compare, on this point, Almas Khan, "Poetic Justice: Slavery, Law, and the (Anti-) Elegiac Form in M. NourbeSe Philip's *Zong!*," *Cambridge Journal of Postcolonial Literary Inquiry* 2, no. 1 (2015): 7.
49  Saidiya V. Hartman, *Lose Your Mother: A Journey Along the Atlantic Slave Route* (New York: Farrar, Straus and Giroux, 2008), 6.
50  Philip, *Zong!*, 184.

51  Ibid., 198.
52  Austen, "*Zong!*'s 'Should we?': Questioning the Ethical Representation of Trauma," 66–67.
53  Philip, *Zong!*, 176. Note, in this context, Philip's statement, "I have often since wondered whether the sounds of those murdered Africans continue to resound and echo underwater. In the bone beds of the sea" (ibid., 203).
54  Howe, *The Birth-mark*, 4.
55  Philip, *Zong!*, 207.
56  See, on Howe's collaboration with the sound artist David Grubbs, Montgomery, *The Poetics of Susan Howe*, 99. See, for recordings of Philip performing parts of *Zong!*: https://www.nourbese.com/zong/.
57  Shannon Ebner, Susan Howe, and Nathaniel Mackey, *STRAY: A GRAPHIC TONE* (Portland, Oregon: Fonograf Editions / Amsterdam: ROMA Publications, 2019), LP.
58  Susan Howe, *Debths* (New York: New Directions, 2017), 9.
59  Howe, "Excerpts from an Interview," Ebner, Howe, and Mackey, *STRAY: A GRAPHIC TONE*.

## Works Cited

Austen, Veronica J. "*Zong!*'s 'Should we?': Questioning the Ethical Representation of Trauma." *ESC* 37, nos. 3–4 (2011): 61–81.

Bernstein, Charles. *Pitch of Poetry*. Chicago: University of Chicago Press, 2016.

Bloomfield, Mandy. *Archaeopoetics: Work, Image, History*. Tuscaloosa: The University of Alabama Press, 2016.

Bruns, Gerald L. "Voices of Construction: On Susan Howe's Poetry and Poetics." *Contemporary Literature* 50, no.1 (2009): 28–53.

Clarke, George Elliott. *Odysseus Home: Mapping African–Canadian Literature*. Toronto: University of Toronto Press, 2002.

Deleuze, Gilles, and Félix Guattari, *A Thousand Plateaus: Capitalism and Schizophrenia*. Translated by Brian Massumi. 1980. London: Continuum, 2004.

Derrida, Jacques. *Specters of Marx: The State of the Debt, the Work of Mourning and the New International*. New York: Routledge, 2006.

DuPlessis, Rachel Blau. "'Whowe': An Essay on Work by Susan Howe." *Sulfur* 20, no.1 (1987): 157–65.

Ebner, Shannon, Susan Howe, and Nathaniel Mackey. *STRAY: A GRAPHIC TONE*. Portland, Oregon: Fonograf Editions / Amsterdam: ROMA Publications, 2019, LP.

Hartman, Saidiya V. *Lose Your Mother: A Journey Along the Atlantic Slave Route*. New York: Farrer, Straus and Giroux, 2008.

Howe, Susan. *Singularities*. Middletown, CT: Wesleyan University Press, 1990.

Howe, Susan. *The Birth-mark: Unsettling the Wilderness in American Literary History*. New York: New Directions, 1993.

Howe, Susan. *My Emily Dickinson*. 1985. New York: New Directions, 2007.

Howe, Susan. *Debths*. New York: New Directions, 2017.

Howe, Susan. "Excerpts from an Interview." In Shannon Ebner, Susan Howe, and Nathaniel Mackey. *STRAY: A GRAPHIC TONE*. Fonograf Editions / ROMA Publications, 2019, LP.

Ikoniadou, Eleni. "The Lament." In *Unsound: Undead*, edited by Steve Goodman, Toby Heys, and Eleni Ikoniadou, 73–5. Falmouth: Urbanomic, 2019.

Kane, Brian. "Sound Studies without Auditory Culture: A Critique of the Ontological Turn." *Sound Studies* 1, no. 1 (2016): 2–21.

Khan, Almas. "Poetic Justice: Slavery, Law, and the (Anti-)Elegiac Form in M. NourbeSe Philip's *Zong!*" *Cambridge Journal of Postcolonial Literary Inquiry* 2, no. 1 (2015): 5–32.

Ma, Ming-Qian. "Articulating the Inarticulate: *Singularities* and the Counter-Method in Susan Howe." *Contemporary Literature* 36, no. 3 (1995): 466–89.

Mackey, Nathaniel. *Discrepant Engagement: Dissonance, Cross-Culturality and Experimental Writing*. 1993. Cambridge: Cambridge University Press, 2009.

Montgomery, Will. *The Poetry of Susan Howe: History, Theology, Authority*. New York: Palgrave Macmillan, 2010.

Morrell, Carol. "Introduction." In *Grammar of Dissent: Poetry and Prose of Claire Harris, M. NourbeSe Philip and Dionne Brand*, edited by Carol Morrell, 9–24. Fredericton: Goose Lane Editions, 1995.

Perloff, Marjorie. "'Collision or Collusion with History': The Narrative Lyric of Susan Howe." *Contemporary Literature* 30, no. 4 (1989): 518–33.

Philip, M. NourbeSe. *Zong!* Middletown, CT: Wesleyan University Press, 2008.

Philip, M. NourbeSe. *She Tries Her Tongue, Her Silence Softly Breaks*. 1989. Middletown, CT: Wesleyan University Press, 2014.

Sharpe, Christina. *In the Wake: On Blackness and Being*. Durham, NC: Duke University Press, 2016.

Staley, Beth Ann. "Emily Dickinson's Echology: A Listener's Reconceptualization of Citizenship, Consciousness, and the World." PhD-Thesis, West Virginia University, 2019.

White, Jenny L. "The Landscapes of Susan Howe's 'Thorow.'" *Contemporary Literature* 47, no. 2 (2006): 236–60.

# 5

# Reframing Indigenous Sonic Archives
## Jeremy Dutcher and the Cultural Politics of Refusal

Sabine Kim

From the 1880s until 1951, Canada sought to repress the singing of songs at Indigenous ceremonies as part of a general ban on cultural practices such as potlatches. Though the culture went underground and often survived, the antipotlatch measures, made enforceable by law via a series of amendments to the Indian Act of 1880, were backed by the threat of fines and imprisonment. Previously the colonial government had sought to prevent the transmission of Indigenous knowledge and the continuation of tradition by relying on individual efforts by priests and other authority figures to discourage participation in the Sun Dance and other ceremonies as part of a so-called civilizing campaign. But with the creation of the Indian Act agents of the state were given new powers to violently assimilate Indigenous peoples. One index of the seriousness with which the settler state viewed Indigenous cultural practice as a threat to its own sovereignty can be found in the response to the potlatch held by 'Namgis Chief Dan Cranmer in Alert Bay, BC in December 1921. Dozens of participants were arrested, twenty-two were sentenced to jail, and others forfeited coppers and hundreds of other precious ceremonial items rather than risk a trial and potential jail term, which often might have been financially devastating.[1] It is no coincidence that assimilative violence became further entrenched at this time, when a federal politics of starvation and the Numbered Treaties had enabled the Canadian state to compel Indigenous peoples to exchange huge swaths of their traditional lands for cramped reserves and hollow promises of education and healthcare.[2] Only the resistance and resilience of Indigenous peoples stood in the way of the powers invested by the new Indian Act, under which Canada was able to appropriate Indigenous territories with impunity. Heidi Kiiwetinepinesiik Stark (Anishinaabe) has given the name "criminal empire" to this dispossessive force of settler colonialism, arguing that the United States and Canada criminalized Indigenous resistance to colonial violence in order to mask the illegality of their own actions.[3] Stark is thinking mainly of settler dispossession of land and title, but here we might also include the confiscation of Alert Bay potlach regalia and ceremonial items, most of which joined the collections of the National Museum of Canada and the Royal Ontario Museum.[4] Though lands and collector items are vastly different entities, they were often acquired in this period through the dual branches of government and public museums, operating on a similar logic of dispossession—likewise with the oral culture of Indigenous nations in general, something which I will examine in this chapter. On the one hand, Indigenous song is a dangerous (because enabling) component of living culture, and therefore settler society desires to repress it as a sonic assertion of Indigenous presence.[5] On the other hand, Indigenous songs, stories, and speech were collected by numerous ethnographers armed with phonographs and working on behalf of museums and/or the state. Although the latter activity seems less repressive, it is continuous with the logic of settler colonialism's dispossessive drive; subject to the museum's organizing taxonomies, the songs are transformed into ethnographic objects that support a settler narrative of history. Placed there for preservation, the recordings are nevertheless separated from their singers and can no longer fulfill their function in upholding the histories and memories in which they are embedded and which they also bring into being, in the moment they are performed. In this chapter, I argue that the different Indigenous and European understandings of ownership mean that the importance of recordings for repatriation work has often been overlooked by museums. What social anthropologist Brian Noble noted in 2002 is still largely true, that "[w]hile museums repatriate bundles, designs, regalia and other *tangible* matter, the vitally important issue of *intangible* matter—as in the example of songs held as recordings in archival collections—remains problematic."[6] To make this argument, I look at composer Jeremy Dutcher (Wolastoqiyik, Maliseet), a member of the Tobique First Nation in New Brunswick, Canada, whose Polaris award–winning album of 2018 can be considered a form of repatriation as it mixes archival Wolastoqey songs with his own songs in the same language.[7] I link this to a larger question of Indigenous sound as museal object in settler archives, and throughout, I consider Dutcher's specific practice in relation to an Indigenous politics of refusal, drawing on Mohawk scholar Audra Simpson and Glen Coulthard (Yellowknives Dene).

## Living Archives: *Wolastoqiyik Lintuwakonawa* and Reframing the Colonial Archive

The concert hall is empty and dim except for the warmly lit stage with the piano. A figure cloaked in a diaphanous, flowing cape and wearing a top hat made out of birchbark with a single, jaunty feather strides out of the darkness and onto the stage. The magician-musician sits down and in an operatic voice begins to sing in the Indigenous language of Wolastoqey and to play the piano.[8] Then, as if growing out of the piano player's body, dancers emerge on the stage. Their dancing is ethereal, phantasmatic. The performance space is suddenly filled with the rich sounds of voice and piano and the beautiful movements of the performers. Though we do not leave this space, we are transported somewhere else. Thus begins the music video of the song "Mehcinut," from Jeremy Dutcher's debut album, *Wolastoqiyik Lintuwakonawa* (2018), which he created using songs belonging to the Wolastoqiyik that had been thought lost for 110 years.[9] Listening to the advice of elder Maggie Paul, Dutcher sought the phonograph recordings made by an American anthropologist who did fieldwork among the

Wolastoqiyik and Maliseet on the East Coast in the first part of the twentieth century and traced them to the Canadian Museum of History in Gatineau, Quebec. There, in an archive, Dutcher encountered a cultural treasure, hearing the voices of his ancestors speaking and singing many songs that had not been heard for years, parts of which he incorporated into an operatic work. The resulting album received much critical and mainstream attention, winning the prestigious Polaris Music Prize in 2018 and the Best Indigenous Music Album of the Year at the 2019 Juno Awards (Canada's version of the Grammys). The video imagines and enacts an Indigenous worldview through a hybrid acoustic song–piano composition–dance performance–visual work that resists the disciplinary ear of the dominant settler society of Canada.[10]

Like other tracks on the album which it opens, "Mehcinut" blends the voice of the composer with those of Wolastoqiyik speakers retrieved from archival recordings of Wolastoq songs and conversation collected by an anthropologist commissioned by a public museum in the first quarter of the twentieth century. This critically acclaimed album has a utopian dimension because it is directed at a language community that has around 360 speakers in Canada and refuses to provide English translation on the album itself or in the liner notes.[11] The Wolastoqey language and by extension the culture are treated as mainstream entities, thus subordinating the hegemonic English and French bicultural landscape of Canada.

*Wolastoqiyik Lintuwakonawa* is significant as a form of counter-archive that reframes the colonial archive. Museum collecting in the nineteenth and early twentieth centuries coincided with an era of intensive imperial expansion accompanied by a consolidation of knowledge—enterprises physically embodied in museums and zoological gardens as showcases of empire and intellectually organized in the emergent disciplines of anthropology, linguistics, geology, archaeology, and zoology. Museums and anthropologists often worked closely together, in the spirit of Franz Boas's designation of museums as material "storehouses" of culture that could be studied by scientists to verify their theories.[12] However, there are certain problems with the collections assembled by the large public museums in this period. One problem is the historic context in which collections were assembled, characterized by Eurocentric paternalism; there is as well a problem of epistemology, in which the objects were categorized inappropriately, in ways that ignored Indigenous law, spirituality, history, culture. Moreover, as Allison Mills (Omushkego) has argued, "because generations of children were taken from the families and forced into residential schools, resulting in an incredible loss of language, culture, and connection" it is a bitter irony that "[i]n Canada and other settler states, these stories may *only* exist in archives, in the sound recordings made by ethnographers."[13]

## Sonic Repatriation: Alternative Conservation outside the Museum

A tension exists between the museum mandate to safeguard items for future generations and the drive to collect, with its privileging of rarity, heightened by anthropologists' self-understanding as adventurers and "hunters."[14] In addition, as the Cranmer potlatch confiscations suggest—six hundred items disappeared into museums across the world, facilitated by a magistrate/Indian agent working with an RCMP sergeant[15]— settler discourse that positions Indigenous elders as criminals for practicing their culture can draw on the same discourse to take the ritual objects into "safekeeping." By drawing a discursive circle around the potlatchers and placing them in a grey area of il/legality while claiming to act within the law itself, settler states reduce Indigenous sovereignty while absorbing the symbolic power of cultural practice into their own national narrative, and the major public museums in the United States and Canada have often benefited from these expressions of colonial power. Settlers feel empowered to coerce First Nations owners, wrongly perceived as remnants of a vanishing culture, into parting with important ceremonial and personal objects.

However, the collectors were not the only ones with agency, and not always the ones who had the upper hand. As Donald Bigknife explained to anthropologist Alison K. Brown, the Piikani Nation in Alberta might not have allowed collectors to view ceremonial objects: "[s]ee, a lot of these old guys, what they had they didn't show them. Special things, sacred things."[16] Only the owner of a medicine bundle is permitted to look inside, and although protocols are culturally specific and even differ from nation to nation, most First Nations have rules for ceremonies and their associated objects that stipulate who is allowed to witness what. Among the Nuxalk of British Colombia, for instance, songs and dances that are associated with owners must have their permission before they can be performed, whether at a potlatch or by students practicing, regardless that this latter takes place outside the context of a ceremony.[17] One of the things that collectors and museums either dismissed or did not understand is this guardianship and the responsibilities that are incurred by taking on ownership of a song, dance, object, or story. Non-Indigenous and Indigenous understandings of ownership differ, as for the Gitanyow of British Columbia:

> Each House holds its unique set of crest images on blankets, rattles, poles and other regalia, and . . . it also holds chief names, songs, and other intangible possessions. In a less fundamental way, the House also holds the actual objects themselves . . . [A] chief's holding of these possessions is not a personal property right; rather, the intangible images, music, and words, as well as their tangible depictions on regalia and poles, are held in trust by the chief and the House members for future generations. The trust property is more than a right to display certain images and to perform certain songs and dramas: it is intimately linked with the people's histories, which constitutionally define each group and its relations with other groups, and it connects each group with its territories.[18]

Such notions of "trust property" that contrast the European understanding of objects as chattel, along with the communal versus individual ownership concept, are not reflected in federal repatriation policy. This difference in definition is one factor that has in the past held back the return of objects to their rightful guardians. Another factor is the value placed on written records in attesting to ownership, versus the visual/oral/aural ties that might bind song, object, singer, and community in Indigenous relational networks. Such ties become hard to trace outside the museum because of cultural loss; when an object is taken away, the song that goes with it also disappears. Inside the museum, these ties are broken by "the separating of songs from bundles, societies from ceremonies, exchange relations from the camp arrangement, rights from all of this, [which] derives largely from the modernist practices of anthropology and museology and museum management."[19] Museums choose to overwrite the Indigenous cultural meaning of a sacred complex of song/object/performance context with a hegemonic perspective when they separate material holdings from sonic objects. At the same time, the repatriation of a material object does not guarantee that the songs associated with it will also be given back (i.e., returned to the owners and deleted from the museum archives), as Reg Crowshoe, a Thunder Medicine Pipe Keeper, experienced in relation to the Blackfoot medicine bundles his nation wished to repatriate in the late 1990s, early 2000s: "This is what the museum people say: 'We'll give you the bundles back . . . But Freedom of Information Act says you can't have the songs because they're on tapes and they're in the archives.'"[20] In a slightly different way, the primacy of the written (in the form of law that has been codified) here again asserts itself over the oral Indigenous form.

In contrast, Dutcher's compositional strategy can be understood as a repatriating act that enables the ancestors' songs to be brought home to the Wolastoqiyik people. It was not an easy task, taking five years from idea to the album's release in 2018, and involving a labor of love in notating the songs over weeks spent in the archives, but it represents a move from the somewhat reified air of the archives to the very public circulation potential afforded by a studio album. Dutcher's transcription is also a reversal of the settler valuing of writing as the form which houses knowledge, rules, and history that was so prominent in anthropological research. Reflecting on the ways Indigenous speech was documented in both recorded and written form by anthropologists working in the late 1890s and early quarter of the twentieth century, Susan Gingell and Wendy Roy identify an "ethnographic overwriting of Indigenous orature"[21] that I would argue could be analyzed in relation to the transcription of song as well as part of a broad anthropological textualization of culture. In the oral–written binary, transcription was assumed to be a neutral act of recording verbal utterances. In fact, it was a reductive process as

> ethnographers who engaged in the massive project of salvage anthropology, powered by the belief that Indigenous peoples were dying out, did fieldwork to collect and write down Indigenous stories. Working from the script-centred view of culture, they thought they were recording the oral texts of primitive cultures when in fact what they were doing was extracting from performances of Indigenous orature. Most of the time, these anthropologists published the results in print form, sometimes—as in Franz Boas's case—as edited or rewritten summaries, thereby creating a large archive of textualized orature.[22]

Similarly, songs in museum archives are separated from the framing provided by the performer or an emcee at a ceremony, who relates the song's origin and introduces the owner associated with the song.

By creating an album sung solely in the Wolastoq language and folding the archival recordings of his ancestors' songs into the new compositions, the composer returns the songs to the Wolastoqiyik people—thus reframing the social life of the acoustic museum object, on Indigenous terms. In this chapter, I will examine what I am calling Dutcher's "living archive" approach. This is directed against museum collections of sound recordings made during the heyday of empire and with mandates to preserve (in all its implications of essentializing gaze and commodifying grasp). Jennifer Lynn Stoever draws attention to Fred Moten and Stephano Harney's distinction between conservation and preservation as a way to think about the latter's conservative, essentializing qualities versus the resistant strategy of conservation, which is innovative and responsive: "[C]onservation is always new. It comes from the place we stopped when we were on the run. It's made from the people who took us in. It's the space they say is wrong, the practice they say needs fixing."[23] Historic ethnographic recordings represent colonial practices of sound collectivity that have all too often viewed Indigenous cultures as being frozen in the past and radically non-contemporaneous. As court cases from the late 1990s onward have suggested (e.g. Peru resorting in 2008 to suing Yale University to repatriate Inca artefacts that should have been returned by agreement in 1916, as well as disputes involving Māori human remains), there is currently an urgent problem of dispossessed Indigenous culture that should/must be repatriated in settler nations such as the United States.[24] Although the problematic aspect of museum ownership of dispossessed "artifacts" has been recently critically examined, less scholarship exists on museum repatriation of sound recordings made by anthropologists and currently housed in the sonic archives of museums.[25] As integral elements of a complex unity of song, ceremony, object, and singer, songs cannot be separated from either the tangible objects to which they give voice or from their ceremonies without a loss of meaning for the whole. Thus, I argue that Indigenous songs must equally be regarded as "cultural goods" that should be reconnected with their rightful owners.

## A "Platform to Share the Truths": Jeremy Dutcher's Album as Counter-Archive

As an Indigenous composer who has used archival phonograph recordings in his opera sung in the Wolastoqey language, Dutcher explicitly frames his work in terms of decolonization and deimperialization. That is, his album is a provocation both regenerative, for the specific and broader First Peoples communities, and also a prompt for the dominant society of Canada's settler colonial state, as he made clear in his acceptance speech for the Polaris Music Prize:

> Canada. You are in the midst of Indigenous Renaissance. Are you ready? To hear the truths that need to be told? Are you ready to see the things that need to be seen? . . . This is incredible . . . Music is changing this land. What you see on the stage tonight, this is the future. This is what's to come . . . To do . . . this record in

my language and have it witnessed not just by my people but people from every nation, from coast to coast up and down Turtle Island. We're at the precipice of something, feels like it.[26]

As a counter-archive, "Mehcinut" addresses the contexts of the original recording, which was undertaken as a Euro-American project of collecting the cultures of what were presumed to be a vanishing people. The Canadian Museum of History's collection of wax cylinder recordings of Wolastoqey songs and commentary was originally gathered by William H. Mechling, an anthropologist who had briefly worked with Franz Boas and did fieldwork with the Passamaquoddy and Wolastoqiyik communities on the East Coast from roughly 1907 to 1913. Stoever's theorization of the sonic color line as enacted by disciplinary listening is relevant here to understand the cultural work that colonial archives do. Noting that "histories of listening are [. . .] not only enmeshed in the matrices of social difference and power but also helping to constitute them," Stoever gives the name "listening ear" to the racializing of listening, in which the terms of legal subjecthood, citizenship, and Americanness itself are guarded by an unspoken whiteness which categorizes whose claims are recognizable (those by subjects who sound white) and whose claims are not (those of "black people, indigenous peoples, immigrants, and colonized peoples").[27] This auditory "discernment" privileging and producing whiteness as a social norm functions in the context of live performance—such as the way that the reception of the Jubilee Singers by nineteenth-century white audiences strove to frame the spirituals within the depoliticizing discourse of plantation nostalgia.[28] In a similar fashion, in creating categories of "folk" and "primitive" music that end up in anthropological collections in contrast to "refined" music belonging to a cultural repertoire, museal practice enacts a sonic color line that erases the complex performance contexts of Indigenous music.

Equally troubling is the assumption, often held by national museums, that museum mandates to preserve culture seem for the general public to have more authority than tribal protocol concerning the manner in which stories may be shared, for whom and by whom. Lastly, the telling of stories is not a unilateral practice but creates a responsibility on the part of the listener to act in the future with the knowledge that he or she has been privy to.[29]

Although museum practices are changing, largely due to the persistence of Indigenous owners of cultural property making specific and repeated demands for repatriation, the museum's listening ear continues to make itself felt insofar that the Wolastoqey recordings were sought out by Dutcher, rather than the other way around. Issues such as limited resources and institutional knowledge that may have been lost as curatorial priorities changed surely make it difficult even for public museums with larger budgets to pursue the question of ownership by a particular community—for example, the Canadian Museum of History has more than 100,000 audiovisual artefacts and more than 3,000 wax cylinder phonograph recordings, of which only about a hundred are related to the Wolastoq tradition. Yet the sonic color line continues to play a role insofar that the Indigenous sonic archive is considered research material and first and foremost understood to belong to an anthropological history. Edward

Sapir's ambition, like Boas's, was to document Indigenous culture for the furthering of knowledge by non-Indigenous society. The recordings may impart knowledge of, for example, Wolastoqiyik culture, yet they are also sound documents that were elements of an imperial project of knowledge production. The importance of the recordings derived in large part from being recorded by an anthropologist for another respected anthropologist, as part of an ambitious project to create a comprehensive archive of Indigenous cultures within a national framework.

## Refusal and the Temporalities of Recorded Sound

With the historic restoration of voices, songs and stories that were lost for more than a hundred years, Dutcher's album defies the framework of atemporality that has historically been imposed by anthropological theories of Indigenous cultures existing in a precultural realm. The album instead insists, as Mohawk anthropologist Audra Simpson has remarked about her people's struggle for self-determination, that "Indigenous people are reminders, sometimes indecipherable *announcements* of other orders, other authorities, and an earlier time that has not fully passed."[30] Moreover, *Wolastoqiyik Lintuwakonawa* is part of a renewed Indigenous cultural and political presence that refuses the accommodationist stance. In 2019, for example, the Grassy Narrows (Asubpeeschoseewagong) First Nation were continuing to press Canada's prime minister to clean up mercury pollution of waterways on their land caused by industrial dumping and to provide health care for the affected community as promised two years earlier; the Idle No More movement has repeatedly called on the Canadian state to respect ancient treaty promises; and there are court cases challenging the authority of Canada to enforce laws when they conflict with Indigenous sovereignty grounded in Indigenous rights on unceded territory. Although ethnicity and race in Canada are often perceived as less politicized in comparison to the United States because, to take one example, there was no Civil Rights Movement, political and social organizing by the Indigenous peoples in Canada were an inspiration, in conjunction with third world decolonizing movements across the globe, for Japanese Canadians, Chinese Canadians, and other racialized groups in the 1970s onward to agitate for social justice.[31] In general, the twenty-first-century defiance by Indigenous people who reject citizenship and refuse to recognize Canada as a political community to which they belong is the dissent of a generation that has come of age. Until 1951, the Indian Act forbade lawyers from advising Indigenous individuals who were Status Indians; in addition, entry into many professions such as law and medicine was de facto blocked for Indigenous persons, who until the 1960s were not able to graduate from university without having to give up their status and rights as Indigenous people.

Audra Simpson's argument about Indigenous resistance as an insistence on an Indigenous political sphere distinct from and outside of the Canadian state helps to excavate the cultural political dimension of Dutcher's compositions and their sounding out of autonomy. In "The Ruse of Consent," Simpson contends that "law in colonial contexts enforced dispossession and then, granted freedom through the legal tricks of consent and citizenship."³² With reference to the Kanehsatà: ke defense of sacred territory threatened by bulldozers in 1990, Simpson writes that the Mohawk "refuse[d] to consent to the apparatus of the state" that first threatened them with death and then expected to set the terms of the subsequent relationship.³³ This refers both to the historic standoff of 1990, in which the state sent the army with tanks to break the Mohawk siege, and to the longer history of violent white dispossession of Native land. The five hundred years of settlement of the Americas have shown that agreements between settler colonialists and Indigenous peoples are interpreted differently by the two sides and that challenges to the white version are responded to with threats of death and actual death; Simpson refers to Heidi Kiiwetinepinesiik Stark's 2016 article, "Criminal Empire: The Making of the Savage in a Lawless Land," in which Stark demonstrates how Indigenous protest was constructed as criminal, an operation of false allegations that simultaneously served the purpose of legitimating the state's claims to sovereignty. Stark highlights the public execution, after a show trial, of six Cree and two Assiniboine leaders of the Frog Lake Resistance (1885) in what is now Manitoba, Canada, noting the government's use of spectacle to send a message intended to intimidate other Indigenous rebels. An earlier public hanging in the United States, of thirty-eight Dakota men in 1862, also was intended to discourage Indigenous resistance.³⁴ In both cases, the executions were the largest to be carried out in either country and reveal the hollowness of the belief that Canadian settlers were not as brutal toward Indigenous peoples as the United States. And thus, Simpson comments, "'[s]ettler time' is revealed as the fiction of the presumed neutrality of time itself, demonstrating the dominance of the present by some over others, and the unequal power to define what matters, who matters, what pasts are alive and when they die."³⁵

*Wolastoqiyik Lintuwakonawa*'s layering of time underscores Dutcher's agency in creating a temporality that defies the ordinary expectations of time and seeks to remember Wolastoqiyik ancestors. Dutcher's mother is a residential school survivor whose memories of being punished for speaking her language are a vivid reminder for Dutcher of the decision that parents in his community took to not speak in the traditional tongue in the presence of children, in an effort to shield them. Remembering and reviving the language that is in danger of extinction is thus an act of refusal to consent to agreeing to vanish, to accept the terms and histories of colonial settlement.

In "Mehcinut," this persistence and defiance is conveyed on different levels. The words of Jim Paul, the Wolastoqiyik elder whose voice is heard retrieved from a time more than a hundred years ago, tell of death and what comes after. The content of the traditional song had been almost forgotten by the time Dutcher listened to it in the archives of the Canadian Museum of History. When Paul's story and chant were recorded between 1907 and 1911, they were part of the living culture of the Tobique First Nation. Just as the process of recording disrupts the linearity of time and offers the chance to bring back something that is no longer existent as such, so too are Jim Paul's words and the Wolastoqiyik belief system that the song conveys kept intact. The content offers a metacommentary on Dutcher's aesthetic project, which symbolically seeks to undo the breaks imposed by colonialism, which led to his people being dispossessed and their traditional lands restricted to a fraction of what they had been, their nation divided artificially by the US-Canada border that the Wolastoq River crosses.

The video version of "Mehcinut," which was published on YouTube in October 2019, differs acoustically in a significant way from the album version.[36] The music video begins with an auditory soundscape that acts like a sonic frame for the song played by the musician. The song is thus embedded within a larger context, and I am arguing that this framing, which is absent in the album version, is a reference to the social worlds that sound creates and from which sound emerges. Sound both embodies and is a condition of this sociality that has been theorized by scholars from a range of disciplines, including linguists (Benveniste, Saussure), cultural theorists (Brooks, Moten, Sterne, Stoever, Weheliye), performance studies scholars (Henriques), and anthropologists (Erlmann).[37] The album version, as discussed earlier, makes this sociality explicit in the combining of the voices that together recreate a shared sociolinguistic space that had been fragmented by the settler vision of culture. The music video extends this living archive in which the songs are carriers of cultural memory into a "live" soundscape" in Nathalie Aghoro's sense of the term that she establishes in her analysis of Karen Tei Yamashita's *I Hotel*, which features the gloriously raucous cacophony of the International Hotel in San Francisco's Manilatown that was a hotbed of activism of all sorts.[38] The urgency of the hotel residents' efforts to prevent their building from being demolished to make way for a development is conveyed by the way the sound presses through the walls of the corridor, infiltrating the rooms and connecting the residents' lives and their individual actions. Pierre Schaeffer's concept of soundscape is wedded here with the notion of sound as something interrelational and inextricably social. Similarly, Jeremy Dutcher's "Mehcinut" can be regarded as a living soundscape because it creates a social world.

The video shifts the orientation of non-Native audiences. The hall looks like an ordinary performance space, where a symphony orchestra might play. Yet in the course of the video, this assumption of familiarity proves to be false. Throughout there is an attentiveness to the existence of the more-than-human. Visual and auditory clues point to the space itself being alive. The first sounds to be heard are an electronic twittering and a burst of voices speaking too quickly to be understood. What we see is a stage bathed in a soft light in a modern concert hall with wooden panels and a backdrop of thin wooden ribbed slats that, in their interplay of shadow and light, evoke a living forest. Dutcher himself wears a birchbark hat with a single feather jauntily stuck in it and a flowing diaphanous cape, black fingernail polish.

The music video and Dutcher's album both partake in what I would call a refusal of normative modes of sounding, looking, and acting; a refusal of the categories and disciplinary practices by which colonial governmentality subtly reproduces its power. This kind of refusal is deeply political, according to Simpson:

> What such requirements involve is a forgetting that the state's very being creates the problems that . . . Indian reserves must manage, and yet, states act as though this is not a matter at all—even though this vexed, very important non-mythical origin story of fundamental *dispossession* is everywhere and nowhere. So, the

implicit demand to forget, through the operation of consent and citizenship, is challenged by the counter that Indigenous people represent simply by (a) living and (b) knowing this. In living and knowing themselves as such, they pose a demand upon the newness of the present, as well as a knotty reminder of something else.[39]

Here again we have Dutcher's double temporality of phonographic recording digitized and inserted into his 2018 album, and a recurring sense of space being occupied in multiple ways. This layering of spatiality and temporality refers, among other things, to the postcolonial space of settler society, where, "in the case of what is now the US, force has been deployed to take land from indigenous peoples and extract labor from black bodies upon that dispossessed land," all the while spinning a narrative of the nation that instantiates the colonial lie of "discovery."[40] The elaborate story of nation that is necessary to mask the dispossession of this proportion, not to mention the accompanying violence, forced assimilation, and genocidal policies of starvation, is undergirded by a reliance on the rule of law, which in turn suggests that the lawfulness of the state means that it itself is a just and fair state. Within this framework whiteness is a possessive: "Whiteness is a form of property that can be used to protect a person from either being rendered as property or having property taken."[41]

The more outright repression and forced assimilation embodied in the residential schools and the series of bans on the performance of various ceremonies, including song and dance, from 1884 to 1951 have given way to what Yellowknives Dene scholar Glen Coulthard terms a colonial politics of recognition in which a liberal politics of multiculturalism weakens the demands for Indigenous self-determination.

## Collecting Sounds and the Politics of Recognition

In *Red Skin, White Masks: Rejecting the Colonial Politics of Recognition* (2014), Coulthard proposes a programmatic critique of the politics of recognition, that is, the state's bestowal of rights and protections for citizens within the purview of its laws, which has characterized Canada's relations with Indigenous peoples. In "ostensibly tolerant, multinational, liberal settler polities such as Canada," the legal framework of recognition masks the continuation of colonialism and the persistence of the oppression of First Peoples.[42]

As Tara Browner (Choctaw) has argued in "Thoughts on Musical Borrowing and the 'Indianist' Movement in American Music" about Euro-American musical scholars and composers who "collected" Indigenous songs during the period of the putatively "vanishing Indian," whether Indigenous music was considered as culture or as precultural acoustic expression, the autonomy of such compositions was ignored.[43] Like the land that was assigned *terra nullius* status because Euro-Americans failed to perceive, or could not imagine land being used in any way other than for productive, extractive capacities, so too were Indigenous nonmaterial cultural goods relegated to a category that conceptually severed the songs from their communities and their traditions. Thus perceived within Euro-American concepts of ownership and authorship, the Indigenous melodies were made available for dispossession by the ethnomusicologists and composers, however well-intentioned they may have believed their collecting efforts to be.

During the early reservation era prior to 1900, Native American songs and music were perceived, Browner argues, "as exotic raw materials that expressed shared universal human emotions, not as products created from a specific musical tradition with its own aesthetics."[44] Despite the generally positive light in which the transcription work of Theodore Baker, the first ethnomusicologist to study Native American music in a comprehensive, pancontinental way, was viewed by his contemporaries for rejecting the "Hiawatha" approach popularized by the poet Longfellow and instead pioneering a "real culture" method, Browner shows that Baker's principle remains problematically mired in a social Darwinist perspective that positions Indigenous culture as a less evolved version of Euro-American culture. In this context, Indigenous music is seen as being universally accessible to Europeans because, according to the theory of cultural evolution formulated by early American anthropologist Lewis Henry Morgan, it represents a stage of cultural development that European composers have progressed beyond. Analyzing the compositions of Edward MacDowell, who was influenced by Baker's scholarship, Browner excavates the philosophical discourse that underpinned this erroneous belief in the under-evolved state of Indigenous music:

> Baker had reinforced MacDowell's beliefs along these lines by matching the Native melodies that he transcribed in his monograph with Greek modes from the period of Aristoxenos (ca 350 B.C.), concluding: "those oldest modes of the Greeks correspond strikingly with those of the North American Indians." Baker's investigation ... set the stage for Western composers to draw upon contemporary Native American music of the late nineteenth century as representations of their own European-derived cultural past.[45]

Within Morgan's conceptual framework, the Indigenous compositions are never Indigenous in their own right but only "Indigenous" on the terms of a colonizing power, in relation to a history of conquest and genocide.

Given the prominence of the phonograph as an anthropological instrument of preservation and documentation, it is worth noting that both Dutcher and Anishinaabe performance and visual artist Rebecca Belmore have used phonographic iconography in their work. The cover of Dutcher's album is a reenactment of a 1910s studio photograph that featured an Indigenous chief, an anthropologist, and a phonograph, but adds a backdrop of a Kent Monkman painting entitled *Teaching the Lost* (2012). Wikipedia describes the original photograph (or a very similar one) as showing Ninna-stako, a South Piegan Blackfeet chief, "listening to a song and interpreting it in Plains Indian Sign Language to ethnomusicologist Frances Densmore."[46] In the original photograph, Densmore and Ninna-stako are pictured on either side of the phonograph. Dutcher's album cover was directed by Monkman (Cree), an artist well known for his queer, ironic versions of settler–Indigenous encounters in which he replaces the figures of the original image with all-Indigenous protagonists. The cover art shows Dutcher seated alone beside the phonograph; on the front, the composer takes the place of Ninna-stako, on the back he sits in Densmore's seat.[47] Taking the dual role of ethnographer and research subject, Dutcher reclaims the power to frame Indigenous sound and performs a refusal of the anthropological totalizing gesture of presuming to capture a culture and own the meaning of its works of art, even as he affirms the value of recording song and music as a way of connecting a people across time and space.

Rebecca Belmore's *Ayum-ee-aawach Oomama-mowan: Speaking to Their Mother* (1991) also repurposes the anthropological recording machine but on an extended collective level.[48] The phonograph is transformed into a giant horn, woven out of wood strips, that needs four people to transport it. The one-way recording system, into which so many traditional Indigenous stories, songs, and even languages themselves seem to disappear, stored away in archives without ready access for the communities that owned them, is reversed in Belmore's activist version. *Ayum-ee-aawach Oomama-mowan* recalls the phonograph in its visual form but is used as a technology of transmission and dialogue rather than of recording and storing.

Belmore's horn derives from a very different understanding of the relationship between speaking, listening, and hearing than the ideology informing the Edisonian phonograph. As Lisa Gitelman has shown, the origins of the American phonograph were closely tied to a Victorian ethos of social improvement, entrepreneurial drive, and scientific ambition in which the "talking machine" would help humans to extend their reach beyond what had hitherto been humanly possible, fixing "fugitive" sounds and creating permanent records.[49] The novelty of sound recording and playback possibilities was something that had to be mediated by the more familiar technology of print media; thus the engravings created at the first demonstrations of phonographs became "souvenir foils" that onlookers would be able to take home: "Those primitive records were clearly meaningful to the women and men who sought them and who were probably asked at the breakfast table the next morning, 'What does it say?' Without the phonograph for playback, the tinfoil records of course said nothing. Yet for the people who brought them home, the very same records clearly said *something*."[50] Gitelman's argument highlights the tensions between recording (storage and archiving) and playback (releasing and replaying) that the phonograph both embodied and bridged. In contrast, the giant horn of *Ayum-ee-aawach Oomama-mowan* is mobile, traveling across the US-Canada border. Meant to be used outdoors, where it can address the land, the horn was created in the wake of the Mohawk resistance to development plans for a golf course on sacred land at Kanehsatà: ke. Elders take turns speaking into the horn, which amplifies their voices but does not capture them. Thus, the horn engages with another form of amplification, namely, of noise, in the sense of Public Enemy's "Bring the Noise,"[51] a voicing that interrupts the status quo and demands action. Though made in response to the Oka standoff, Belmore has since toured the work to other Indigenous communities where Indigenous activists are resisting resource extraction development.

Musical collecting of Indigenous songs may seem benign but has to be regarded as dispossession by another name. Hopi anthropologist Wendy Rose has warned of the dangers of loss of cultural control: "The anthropologist or folklorist hears a story or song and ... reproduces it, eventually catalogs it, and perhaps publishes it. According to the culture of the scholar, it is now owned by 'science.' It has been stolen as surely as if it had been a tangible object removed by force."[52] The self-obviousness of this logic of preservation has begun to be revised in the past two decades, as Indigenous peoples around the world have successfully pursued repatriation of cultural goods.[53] Dutcher raises the seriousness of this issue on the track "Eqpahak," in one of the few instances of English being spoken on the album; the content is pointedly intended to be heard by a settler audience as well. The song opens with Wolastoqiyik elder Maggie Paul in conversation with Dutcher, explaining the interrelated role of song, dance, and ceremony as carriers of intangible culture. She is not the only one bringing the songs back: "[T]here's a lot of people bringing the songs back. When you bring the songs back, you bring the dances back. You're going to bring the people back. You'll bring everything back."[54] The unspoken subtext is the genocidal repression of Indigenous culture expressed in the ban on ceremonies and the attempted enforcement of assimilation in residential schools, where the speaking of Indigenous languages was forbidden and the breaking of ties to family, place, and community was encouraged. In his comparison of First Nations–run museums and the large public museum, James Clifford's critique, posed in 1997, has become ever-more crucial: "The idea of majority institutions such as the Canadian Museum of Civilization and the Museum of the American Indian representing Native American cultures to the nation as a whole is increasingly questionable. So is the very existence of elaborate, enormously valuable, noncirculating collections."[55] It is in this context that we have to understand Dutcher's conception of his role as a medium for the conservation of culture and as a channel for speaking truth to power:

> I feel an immense sense of responsibility to do this work and to have a platform and a spotlight to share truths ... I hope to make that a reality ... I hope to step into that role as a truth teller to share music because I feel like actually it's in beauty that we move forward ... when we change hearts it comes in beauty. I take the responsibility of being a conveyer of beauty very seriously.[56]

## Notes

1  In total, around six hundred items were taken. The coercive nature of the "agreements" has been identified by elders who remember the arrests and have spoken to the 'Namgis Nation's legal counsel as part of the pursuit of repatriation. The Indian agent for Alert Bay arranged to amend the antipotlatch act so that offences would be summary convictions, allowing him to preside over the cases in his dual role as agent and magistrate, and thereby being in a position to arrange the confiscation of regalia in exchange for lenient sentencing (Catherine Bell, Heather Raven, and Heather McQuaig, "Recovering from Colonization," in *First Nations Cultural Heritage and Law: Case Studies, Voices, and Perspectives*, ed. Catherine Bell and Val Napoleon [Vancouver: University of British Columbia Press, 2008],

53–4). This agent, William Halliday, also arranged to ship objects to museums and collectors: "The distance between repression and appropriation was never so slim as it was in the Cranmer potlatch prosecutions" (Ronald W. Hawker, *Tales of Ghosts: First Nations Art in British Columbia, 1922-61* [Vancouver: University of British Columbia Press, 2003], 18). For more on the confiscation as well as Kwagiulth efforts to repatriate the objects, see Harry Assu (Lekwiltok), with Joy Inglis, "Renewal of the Potlatch at Cape Mudge," in *Assu of Cape Mudge: Recollections of a Coastal Indian Chief* (Vancouver: University of British Columbia Press, 1989), 103–8; see also Daisy Sewid-Smith, *Prosecution or Persecution* (Cape Mudge, BC: Nu-yum-balees Society, 1979); for details on the DIA involvement, see Hawker, *Tales of Ghosts*, 22–4.

2 In contrast to the earlier nation-to-nation phase of treaty-making that aimed at European-Indigenous alliances; see Cole Harris, with cartography by Eric Leinberger, *Making Native Space: Colonialism, Resistance and Reserves in British Columbia* (Vancouver: University of British Columbia Press, 2002).

3 Heidi Kiiwetinepinesiik Stark, "Criminal Empire: The Making of the Savage in a Lawless Land," *Theory & Event* 19, no. 4 (2016): n.p.

4 US collector George Heye bought thirty-three pieces for the Smithsonian National Museum of the American Indian and the British Museum (in "a worse irony," these regalia items ended up in the NMAI's "cavernous Bronx warehouse," which James Clifford describes as "bursting with Native American artifacts, most of which have never been, and may never be, displayed"; see James Clifford, *Routes: Travel and Translation in the Late Twentieth Century* [Cambridge, MA: Harvard University Press, 1997], 139). Halliday's boss, Duncan Campbell Scott, arranged for him to send certain masks for Scott's personal collection, to be displayed outside Scott's office (Hawker, *Tales of Ghosts*, 183). Scott, who was deputy superintendent general of the Department of Indian Affairs between 1913 and 1932, embodies a variant form of criminal empire since he wrote poems in the style of Longfellow, lamenting the decline of the Indigenous peoples whose dire conditions were a result of policies he oversaw; the DIA's blatant inaction in the face of massively disproportionate rates of infectious illness and death at the residential boarding schools forms part of the justification for a recent call for Canada's treatment of Indigenous peoples to be named genocide (Phil Fontaine and Bernie Farber, "What Canada Committed against First Nations Was Genocide: The UN Should Recognize It," *The Globe and Mail*, October 14, 2013; see also James Daschuk, *Clearing the Plains: Disease, Politics of Starvation, and the Loss of Aboriginal Life* (Regina, SK: University of Regina Press, 2013).

5 Patrick Wolfe, "Settler Colonialism: The Elimination of the Native," *Journal of Geocide Research* 8, no. 4 (2006): 387.

6 Brian Noble, "Niitooii—'The Same That Is Real': Parallel Practice, Museums, and the Repatriation of Piikani Customary Authority," *Anthropologica* 44 (2002): 123.

7 Looking at a recent trend in Canada toward collaboration between Indigenous and non-Indigenous musicians and composers and drawing on Eva Mackey, *The House of Difference: Cultural Politics and National Identity in Canada* (Toronto: University of Toronto Press, 2002), Dylan Robinson (Stó:lō) has analyzed the accommodationist tendencies of settler music culture, which he criticizes for mistaking formal innovation for broadscale structural change within the economic, social, and political spheres (Dylan Robinson, *Hungry Listening: Resonant Theory for Indigenous Sound Studies* [Minneapolis: University of Minnesota Press, 2020]).

8   Wolastoqey is the traditional language of the Wolastoqiyik, whose lands straddle the border between Maine, United States, and New Brunswick, Canada, along the Wolastoq River. The Wolastoqiyik were among those whose ancestral lands were dispossessed through the creation of reserves in the nineteenth century. The consequences of the removal on their ability to continue their former self-sufficient and self-determined lives continue to reverberate today, with land claims such as the Madawaska Maliseet Canadian Pacific Railway Specific Claim (settled 2008) or those involving the 1892 Tobique surrender of lands being brought to court today (decided September 2016). See Tom McFeat, with Michelle Felice, "Wolastoqiyik (Maliseet)," *Canadian Encyclopedia*, first published 2006 and updated October 2018.
9   Jeremy Dutcher, "Mehcinut: Official Video," released 2019, video, 5:47, https://www.youtube.com/watch?v=6pDRpDjrBZE.
10  I am taking the concept of the disciplinary listening ear, which I discuss in more detail below, from Jennifer Lynn Stoever, *The Sonic Color Line: Race and the Cultural Politics of Listening* (New York: New York University Press, 2016).
11  According to the 2016 census in Canada (McFeat, "Wolastoq," n. p.).
12  For a critical analysis of this, see Alison K. Brown, *First Nations, Museums, Narrations: Stories of the Franklin Motor Expedition to the Canadian Prairies* (Vancouver: University of British Columbia Press, 2014), 64.
13  Allison Mills, "Learning to Listen: Archival Sound Recordings and Indigenous Cultural and Intellectual Property," *Archivaria* 83 (2017): 113.
14  On the metaphor of hunting, see Brown, *First Nations, Museums, Narrations*, 136, see also 133–54. James Clifford's chapter "Four Northwest Museums" remains relevant for insights into the role of museums in the collecting craze over objects from the Northwest Coast especially; see Clifford, *Routes*, 107–46.
15  Kevin Griffin, "This Week in History, 1921: Mass Arrests at Kwakwaka'wakw Potlatch Took Place Christmas Day," *Vancouver Sun*, December 23, 2016, https://vancouversun.com/news/local-news/this-week-in-history-1921-mass-arrests-at-kwakwakawakw-potlatch-took-place-christmas-day/.
16  Qtd. in Brown, *First Nations, Museums, Narrations*, 133, epigraph.
17  Jennifer Kramer, *Switchbacks: Art, Ownership, and Nuxalk National Identity* (Vancouver: University of British Columbia Press, 2006), 74–5.
18  Qtd. in Robert G. Howell and Roch Ripley, "The Interconnection of Intellectual Property and Cultural Property (Traditional Knowledge)," in *Protection of First Nations Cultural Heritage: Laws, Policy, and Reform*, ed. Catherine Bell and Roger K. Paterson (Vancouver: University of British Columbia Press, 2009), 227.
19  Noble, "Niitooii—'The Same That Is Real,'" 123.
20  Qtd. in Noble, "Niitooii—'The Same That is Real,'" 123.
21  Susan Gingell and Wendy Roy, "Opening the Door to Transdisciplinary, Multimodal Communication," in *Listening Up, Writing Down, and Looking Beyond: Interfaces of the Oral, Written, and Visual*, ed. Susan Gingell and Wendy Roy (Waterloo, ON: Wilfrid Laurier University Press, 2012), 26.
22  Gingell and Roy, "Opening the Door," 25.
23  Moten and Harney, *Undercommons*, 63; qtd. in Stoever, *The Sonic Color Line*, 146.
24  For details of Peru's lawsuit and the history of the Machu Picchu treasures, see "Peru Sues" on the Lexis-Nexis database. The NPR Radio website explains the agreement reached out of court in 2010, "Yale Returns Machu Picchu Artifacts to Peru." Māori iwi working together with the government of Aotearoa/New Zealand

and Te Papa Museum to demand the return of toi moko, kōiwi tangata and koimi tangata have been at the forefront of repatriation activism around the world (*Te Papa Tongarewa: Museum of New Zealand*, https://www.tepapa.govt.nz/about/repatriation). Repatriation is entangled with postcolonial histories, as discussed in relation to France's colonial presence in sub-Saharan Africa and the subsequent issue of repatriation of cultural objects in French museums by Felwine Sarr and Bénédicte Savoy's 2018 report, *Rapport sur la restitution du patrimoine culturel africain. Vers une nouvelle éthique relationnelle* (http://restitutionreport2018.com/); with thanks to Nathalie Aghoro for drawing my attention to this. For a critical analysis of museum practices, see Michael Bachmann, "Ambivalent Pasts: Colonial History and the Theatricalities of Ethnographic Display," *Theatre Journal* 69, no. 3 (2017): 299–319.

25 See for example Catherine Bell and Val Napoleon, eds., *First Nations Cultural Heritage and Law: Case Studies, Voices and Perspectives* (Vancouver: University of British Columbia Press, 2008); Catherine Bell and Roger K. Paterson, eds., *Protection of First Nations Cultural Heritage: Laws, Policy, and Reform* (Vancouver: University of British Columbia Press, 2009); Cressida Fforde, C. Timothy McKeown, and Honor Keeler (Cherokee), eds., *The Routledge Companion to Indigenous Repatriation: Return, Reconcile, Renew* (Oxford and New York: Routledge, 2020); and Joe Edward Watkins, *Sacred Sites and Repatriation* (Broomall, PA: Chelsea House Publishers, 2009).

26 "2018 Polaris Music Prize Winner!" released September 18, 2018, video, 8:53, https://www.youtube.com/watch?v=qIEHxNGJApA.

27 Stoever, *The Sonic Color Line*, 19, 32.

28 Stoever stresses the agency that the Jubilee Singers enacted in resisting the listening ear of white audiences. Drawing on original archival research, she demonstrates that the Singers created new versions of slave songs that deliberately estranged the white listening ear, "sounds that the listening ear declared 'need[ed] fixing' " (Stoever, *The Sonic Color Line*, 146).

29 See Thomas King, *The Truth About Stories* (Minneapolis: University of Minnesota Press, 2008).

30 Audra Simpson, "The Ruse of Consent and the Anatomy of 'Refusal': Cases from Indigenous North America and Australia," *Postcolonial Studies* 20, no. 1 (2017): 22.

31 For the argument concerning the role of Indigenous activism in shifting the discourse on race and rights in Canada, see Iyko Day's "Must all Asianness be American? The Census, Racial Classification, and Asian Canadian Emergence," *Canadian Literature* 199 (2008): 45–70.

32 Simpson, "The Ruse of Consent," 20, 10.1080/13688790.2017.1334283.

33 Ibid., 22.

34 Her article highlights the two mass hangings but she notes there were numerous hangings of Indigenous resisters that took place around this time of Modoc, Tlingit, and Nisqually leaders in the United States and Tsliquotin, Ojibwe and Métis leaders north of the border. Stark, "Criminal Empire," n. pag.

35 Simpson, "The Ruse of Consent," 22.

36 See Jeremy Dutcher, "Mehcinut," recorded 2018, track 1 on *Wolastoqiyik Lintuwakonawa*, Fontana North, compact disc.

37 Emile Benveniste, *Problems in General Linguistics*, trans. Mary Elizabeth Meek (Coral Gables, FL: University of Miami Press, [1968–74] 1971); Ferdinand de Saussure, *Course in General Linguistics*, ed. Charles Bally and Albert Sechehaye,

trans. Wade Baskin, introd. Jonathan Culler (Glasgow: Fontanta, [1916] 1981); Daphne A. Brooks, "'Sister, Can You Line It Out?': Zora Neale Hurston and the Sound of Angular Black Womanhood," *Amerikastudien / American Studies* 55, no. 4 (2010): 617–27; Fred Moten, *In the Break: The Aesthetics of the Black Radical Tradition* (Minneapolis: University of Minnesota Press, 2003); Jonathan Sterne, *The Audible Past: Cultural Origins of Sound Reproduction* (Durham, NC: Duke University Press, 2003); Jennifer Lynn Stoever, " 'Doing Fifty-five in a Fifty-four': Hip Hop, Cop Voice and the Cadence of White Supremacy in the United States," *Journal of Interdisciplinary Voice Studies* 3, no. 2 (2018): 115–31; Stoever, *The Sonic Color Line*; Alexander G. Weheliye, *Phonographies: Grooves in Sonic Afro-Modernity* (London and Durham, NC: Duke University Press, 2005); Julian Henriques, *Sonic Bodies: Reggae Sound Systems, Performance Techniques, and Ways of Knowing* (London: Continuum, 2011); and Veit Erlmann, *Nightsong: Power, Performance, and Practice in South Africa*, introd. Joseph Bekhizizwe Shabalala (Chicago: University of Chicago Press, 1995).

38 Nathalie Aghoro, *Sounding the Novel: Voice in Twenty-First Century American Fiction* (Heidelberg: Universitätsverlag Winter, 2019), 183.
39 Simpson, "The Ruse of Consent," 22.
40 Audra Simpson, "Review of *The White Possessive: Property, Power, and Indigenous Sovereignty* by Aileen Moreton-Robinson," *Anthropological Quarterly* 89, no. 4 (2016): 1306.
41 Ibid., 1308. See historian James Daschuk, *Clearing the Plains*, for an account of the government policy of ethnocide in Canada in the late nineteenth century.
42 Glen Sean Coulthard, *Red Skin, White Masks: Rejecting the Colonial Politics of Recognition* (Minneapolis: University of Minnesota Press, 2014), 15.
43 Tara Browner, "'Breathing the Indian Spirit': Thoughts on Musical Borrowing and the 'Indianist' Movement in American Music," *American Music* 15, no. 3 (1997): 265–84.
44 Ibid., 270.
45 Ibid., 271.
46 "Mountain Chief," *Wikipedia* English version, caption for sidebar image.
47 For both front and back cover, Dutcher is posed against a backdrop of a Kent Monkman painting, *Teaching the Lost*.
48 Rebecca Belmore, *Ayum-ee-aawach Oomama-mowan: Speaking to Their Mother*, sound sculpture, made of wood, megaphone (1991, 1992, 1996, and other dates).
49 Lisa Gitelman, "Souvenir Foils: On the Status of Print at the Origin of Recorded Sound," in *New Media, 1740–1915*, ed. Lisa Gitelman and Geoffrey B. Pingree (Cambridge, MA: MIT Press, 2003), 157–73.
50 Ibid., 158.
51 Public Enemy, "Bring the Noise," released 1988, track 2 on *It Takes A Nation of Millions To Hold Us Back*, Def Jam Recordings, vinyl LP.
52 Qtd. in Browner, "'Breathing the Indian Spirit,'" 281.
53 See Sabine Kim, "Traveling Totems and Networks of Mobility: Indigenous Challenges to Dispossession," in *Comparative Indigenous Studies*, ed. Mita Banerjee (Heidelberg: Universitätsverlag Winter, 2016), 41–56.
54 Jeremy Dutcher, "Eqpahak," recorded 2018, track 3 on *Wolastoqiyik Lintuwakonawa*, Fontana North, compact disc.
55 Clifford, *Routes*, 139.

56 "Arkells Invite Jeremy Dutcher onto Stage to Finish Acceptance Speech," *CBC News*, released March 17, 2019, video, 9:35, https://www.youtube.com/watch?v=dFVzqDfiHuo.

## Works Cited

"2018 Polaris Music Prize Winner!" Released September 18, 2018. Video, 8:53. https://www.youtube.com/watch?v=qIEHxNGJApA.
Aghoro, Nathalie. *Sounding the Novel: Voice in Twenty-First Century American Fiction*. Heidelberg: Universitätsverlag Winter, 2019.
Assu, Harry, with Joy Inglis. *Assu of Cape Mudge: Recollections of a Coastal Indian Chief*, 103–8. Vancouver: University of British Columbia Press, 1989.
Bachmann, Michael. "Ambivalent Pasts: Colonial History and the Theatricalities of Ethnographic Display." *Theatre Journal* 69, no. 3 (2017): 299–319.
Bell, Catherine, Heather Raven, and Heather McQuaig. "Recovering from Colonization." In *First Nations Cultural Heritage and Law: Case Studies, Voices, and Perspectives*, edited by Catherine Bell and Val Napoleon, 33–91. Vancouver: University of British Columbia Press, 2008.
Bell, Catherine, and Val Napoleon, eds. *First Nations Cultural Heritage and Law: Case Studies, Voices, and Perspectives*. Vancouver: University of British Columbia Press, 2008.
Bell, Catherine, and Roger K. Paterson, eds. *Protection of First Nations Cultural Heritage: Laws, Policy, and Reform*. Vancouver: University of British Columbia Press, 2009.
Belmore, Rebecca. *Ayum-ee-aawach Oomama-mowan: Speaking to Their Mother*. Sound sculpture installation, 1991; various locations 1991, 1992, 1996.
Benveniste, Emile. *Problems in General Linguistics*. Translated by Mary Elizabeth Meek. Coral Gables, FL: University of Miami Press, 1971.
Brooks, Daphne A. "'Sister, Can You Line It Out?': Zora Neale Hurston and the Sound of Angular Black Womanhood." *Amerikastudien / American Studies* 55, no. 4 (2010): 617–27.
Brown, Alison K. *First Nations, Museums, Narrations: Stories of the Franklin Motor Expedition to the Canadian Prairies*. Vancouver: University of British Columbia Press, 2014.
Browner, Tara. "'Breathing the Indian Spirit': Thoughts on Musical Borrowing and the 'Indianist' Movement in American Music." *American Music* 15, no. 3 (1997): 265–84. JSTOR.
CBC News. "Arkells Invite Jeremy Dutcher onto Stage to Finish Acceptance Speech." Released March 17, 2019. Video, 9:35. https://www.youtube.com/watch?v=dFVzqDfiHuo.
Clifford, James. *Routes: Travel and Translation in the Late Twentieth Century*. Cambridge, MA: Harvard University Press, 1997.
Coulthard, Glen Sean. *Red Skin, White Masks: Rejecting the Colonial Politics of Recognition*. Minneapolis: University of Minnesota Press, 2014.
Daschuk, James. *Clearing the Plains: Disease, Politics of Starvation, and the Loss of Aboriginal Life*. Regina, SK: University of Regina Press, 2013.
Day, Iyko. "Must all Asianness be American? The Census, Racial Classification, and Asian Canadian Emergence." *Canadian Literature* 199 (2008): 45–70.

Dutcher, Jeremy. "Eqpahak." Recorded 2018. Track 3 on *Wolastoqiyik Lintuwakonawa*. Fontana North, compact disc.
Dutcher, Jeremy. "Mehcinut." Recorded 2018. Track 1 on *Wolastoqiyik Lintuwakonawa*. Fontana North, compact disc.
Dutcher, Jeremy. "Mehcinut: Official Video." Released 2019. Video, 5:47. https://www.youtube.com/watch?v=6pDRpDjrBZE.
Erlmann, Veit. *Nightsong: Power, Performance, and Practice in South Africa*. Introduced by Joseph Bekhizizwe Shabalala. Chicago: University of Chicago Press, 1995.
Fforde, Cressida, C. Timothy McKeown, and Honor Keeler, eds. *The Routledge Companion to Indigenous Repatriation: Return, Reconcile, Renew*. Oxford and New York: Routledge, 2020.
Fontaine, Phil, and Bernie Farber. "What Canada Committed against First Nations was Genocide: The UN should Recognize It." *The Globe and Mail*, October 14, 2013. https://www.theglobeandmail.com/opinion/what-canada-committed-against-first-nations-was-genocide-the-un-should-recognize-it/article14853747/.
Gingell, Susan, and Wendy Roy. "Opening the Door to Transdisciplinary, Multimodal Communication." In *Listening Up, Writing Down, and Looking Beyond: Interfaces of the Oral, Written, and Visual*, edited by Susan Gingell and Wendy Roy, 1–53. Waterloo, ON: Wilfrid Laurier University Press, 2012.
Gitelman, Lisa. "Souvenir Foils: On the Status of Print at the Origin of Recorded Sound." In *New Media, 1740–1915*, edited by Lisa Gitelman and Geoffrey B. Pingree, 157–73. Cambridge, MA: MIT Press, 2003.
Greene, Sarah. "Jeremy Dutcher's Innovative 'Wolastoqiyik Lintuwakonawa' Is Really About the Future." *Exclaim!*, April 28, 2018. http://exclaim.ca/music/article/jeremy_dutchers_innovative_wolastoqiyik_lintuwakonawa_is_really_about_the_future.
Griffin, Kevin. "This Week in History, 1921: Mass Arrests at Kwakwaka'wakw Potlatch Took Place Christmas Day." *Vancouver Sun*, December 23, 2016. https://vancouversun.com/news/local-news/this-week-in-history-1921-mass-arrests-at-kwakwakawakw-potlatch-took-place-christmas-day/.
Harris, Cole, with cartography by Eric Leinberger. *Making Native Space: Colonialism, Resistance and Reserves in British Columbia*. Vancouver: University of British Columbia Press, 2002.
Hawker, Ronald W. *Tales of Ghosts: First Nations Art in British Columbia, 1922–61*. Vancouver: University of British Columbia Press, 2003.
Henriques, Julian. *Sonic Bodies: Reggae Sound Systems, Performance Techniques, and Ways of Knowing*. London: Continuum, 2011.
Howell, Robert G., and Roch Ripley. "The Interconnection of Intellectual Property and Cultural Property (Traditional Knowledge)." In *Protection of First Nations Cultural Heritage: Laws, Policy, and Reform*, edited by Catherine Bell and Roger K. Paterson, 223–47. Vancouver: University of British Columbia Press, 2009.
Kim, Sabine. "Travelling Totems and Networks of Mobility: Indigenous Challenges to Dispossession." In *Comparative Indigenous Studies*, edited by Mita Banerjee, 41–56. Heidelberg: Universitätsverlag Winter, 2016.
King, Thomas. *The Truth about Stories: A Native Narrative*. Minneapolis: University of Minnesota Press, 2008.
Kramer, Jennifer. *Switchbacks: Art, Ownership, and Nuxalk National Identity*. Vancouver: University of British Columbia Press, 2006.
Mackey, Eva. *The House of Difference: Cultural Politics and National Identity in Canada*. Toronto: University of Toronto Press, 2002.

McFeat, Tom, with update by Michelle Felice. "Wolastoqiyik (Maliseet)." *Canadian Encyclopedia*. First published 2006. Updated October 2018.

Mills, Allison. "Learning to Listen: Archival Sound Recordings and Indigenous Cultural and Intellectual Property." *Archivaria* 83 (2017): 109–24. https://archivaria.ca/index.php/archivaria/article/view/13602.

Moten, Fred. *In the Break: The Aesthetics of the Black Radical Tradition*. Minneapolis: University of Minnesota Press, 2003.

*NPR Radio*. "Yale Returns Machu Picchu Artifacts to Peru." *All Things Considered*, December 15, 2010. https://www.npr.org/2010/12/15/132083890/yale-returns-machu-picchu-artifacts-to-peru?t=1575709941838.

Noble, Brian. "Niitooii—'The Same That Is Real': Parallel Practice, Museums, and the Repatriation of Piikani Customary Authority." *Anthropologica* 44 (2002): 113–30.

"Peru Sues Yale for Return of Incan Artifacts." *Law360*, December 8, 2008. https://www.law360.com/articles/79618/peru-sues-yale-for-return-of-incan-artifacts.

Public Enemy. "Bring the Noise." Released 1988. Track 2 on *It Takes A Nation of Millions To Hold Us Back*. Def Jam Recordings, vinyl LP.

Robinson, Dylan. *Hungry Listening: Resonant Theory for Indigenous Sound Studies*. Minneapolis: University of Minnesota Press, 2020.

Sarr, Felwine, and Bénédicte Savoy. *Rapport sur la restitution du patrimoine culturel africain. Vers une nouvelle éthique relationnelle*. November 29, 2018. http://restitutionreport2018.com/.

Saussure, Ferdinand de. *Course in General Linguistics*, edited by Charles Bally and Albert Sechehaye, translated by Wade Baskin, introduced by Jonathan Culler. Glasgow: Fontana, 1981.

Sewid-Smith, Daisy. *Prosecution or Persecution*. Cape Mudge: Nu-yum-balees Society, 1979.

Simpson, Audra. "The Ruse of Consent and the Anatomy of 'Refusal': Cases from Indigenous North America and Australia." *Postcolonial Studies* 20, no. 1 (2017): 18–33.

Simpson, Audra. "Review of *The White Possessive: Property, Power, and Indigenous Sovereignty* by Aileen Moreton-Robinson." *Anthropological Quarterly* 89, no. 4 (2016): 1305–10.

Stark, Heidi Kiiwetinepinesiik. "Criminal Empire: The Making of the Savage in a Lawless Land." *Theory & Event* 19, no. 4 (2016): n.p.

Sterne, Jonathan. *The Audible Past: Cultural Origins of Sound Reproduction*. Durham, NC: Duke University Press, 2003.

Stoever, Jennifer Lynn. *The Sonic Color Line: Race and the Cultural Politics of Listening*. New York: New York University Press, 2016.

Stoever, Jennifer Lynn. "'Doing Fifty-five in a Fifty-four': Hip Hop, Cop Voice and the Cadence of White Supremacy in the United States." *Journal of Interdisciplinary Voice Studies* 3, no. 2 (2018): 115–31.

Watkins, Joe Edward. *Sacred Sites and Repatriation*. Broomall, PA: Chelsea House Publishers, 2009.

Weheliye, Alexander G. *Phonographies: Grooves in Sonic Afro-Modernity*. London and Durham, NC: Duke University Press, 2005.

Wolfe, Patrick. "Settler Colonialism: The Elimination of the Native." *Journal of Genocide Research* 8, no. 4 (2006): 387–409.

# 6

# Unsettled Scores

## Listening to Black Oklahoma on the American "Frontier"

Tsitsi Jaji

Until relatively recently in human history, sound was rarely a solitary experience. Listening, rather, has long implied a multiplicity of sorts. At its most basic (and admittedly ablelist) to listen was to integrate the minutely syncopated signals in two ears into an encounter with the sonic. But this chapter takes up the promise of such a scene, a site where many bodies find themselves in proximity, all unable to blink their ears or turn away from a sonic imprint. In other words, they listen together. I propose this is a site of potential solidarity, a moment when, jointly subject to an involuntary bodily experience and its attendant impulses toward interpretation, the many might imagine themselves as one. Hearing together becomes listening in stereo, a joint act of accommodating differences in perception out of a shared act of sensation. This, as I have argued more fully elsewhere, is what the stereophonic can teach us of solidarity.[1] In what follows I pursue this idea by asking how does a collocation coalesce, how does a crowd become a gathering, in other words under what circumstances might co-presence in space and simultaneity in time weave enough of a shared consciousness to claim a collective imagination, however ephemeral? This chapter will take up these questions waiting in two distinct spaces: a site-specific installation and a recording studio.

We begin with Senegalese sculptor Ousmane Huchard Sow's installation, *The Battle of Little Big Horn*.[2] The work comprises scores of larger-than-life figures—Native American and US armed forces, civilian Native Americans, horses, and livestock—recreating the scene of an 1876 battle in the Montana Territory where Crazy Horse led his fellow Sioux to an astonishing victory over General Custer and his Seventh Cavalry division—a rare instance of successful resistance to violent colonization. The impulse toward representing indigenous Americans in struggle echoes in the languages of colonization on the African continent. The French used the term *indigène* to designate Black Africans as outside the bounds of whiteness and civilization, and to allow for colonial education to transform *evolués* into subjects. "Native" was deployed similarly in British colonies. These terms, central to the parallel colonizing projects on both sides of the Atlantic, highlight an incipient solidarity that Black artists activate when engaging with settler colonialism in the Western hemisphere. Recalling in our inner ear the reverberation of these vocabularies prompts us to take coeval conquer as an acoustic structure. Listening to history, we hear a near unison of territorial dispossession at the 1884–5 Berlin Conference for Africa, and in the 1887 Dawes Act for Native America, which authorized the president of the United States to survey American Indian tribal land and divide it into allotments for individual Indians. The sale of "excess" Indian tribal lands to non-Natives under the Dawes Act paralleled the impact of the Berlin Conference, effecting the rapid transfer of land out of Native hands so that by 1934, the share went from 138 million acres to 48 million.[3] The grounds for solidarity are quite literally, then, the ground lost to imperialist seizure. Sow's installation, *The Battle of Little Bighorn*, stands on this ground, a monument to the heroic courage of indigenous peoples so often imagined to have been completely annihilated. But what makes this work relevant to the questions at hand?

The rough sculptural textures emerge from Sow's improvisational techniques, working without sketches or models, build out from metal bars to layers of jute covered with a range of materials including "clay, plastic, glue, stone, metal, fabric, wood, plaster, rubber and random 'found' objects"—in other words, both construction and exhibition are instances of a time-based art work.[4] All these are the sounds of *the Battle of Little Bighorn*, a work distinguished by its massive scale, mobility, and acoustical range: It is haunted by the lost echoes of indigenous American resistance to the ever-westward push of settler colonialism. The installation was first mounted in January 1999 on the shores of the Atlantic in Dakar, a few kilometers from Sow's *atelier*. Soon after, it traveled to France, where it was exhibited for two months on the bridge known as the *Pont des Arts* in Paris.

If, at first glance, the work appears to be static, material, and silent, this is only the case if we take the finished installation as a self-contained work. However, one cannot close one's ears. Just as a musical work is not merely a score, but its multiple performances, made unique by its performers, listeners, and the acoustics of each sounding occasion from rehearsal room to stage, so too Sow's installation is a material repository of an extended sonic sequence—from the clang of tools constructing sculptural frames, to the low hum of artist and apprentices in conversation, to the sounds of breaking waves, wind, sand, and passersby on the beaches of Gorée. This location, on the island off the coast of Dakar where countless captive Africans were held and launched on their journey across the ocean and into slavery, is significant, especially given that the sculptures were installed on the site of a future memorial. The sounds of the port, muffled by the walls of the shipping container, reverberate in sinister tones in light of this history. Arriving at its second site, the sculptures become audiovisual installations incorporating the Paris traffic en route to the *Pont des Arts*, the exclamations and voiced opinions about the artwork—or continuing unrelated conversations—weaving in and out of the exhibition space, a bridge funded by the municipal government of Paris.

By rendering this cluster of material art objects in Senegal and then transporting the works to France, Sow inscribes onto a Parisian landscape where history's silences go unnoticed the brute force of colonial memory writ large, encompassing the violence of American settler colonialism and, by extension, the French metropole's legacy in Africa and beyond. To experience the installation is not merely a visual act, but an *audiovisual* one, a transit around and through objects in open space, pervious to the surrounding ambient sounds of nature and human presence. The noisy echoes of a battlefield evoked with the disruptive scale and African materials of Sow's sculptures bring passers-by into a visceral relation with history that is not only seen but heard in stereo with the present. The open-air exhibition illustrates the classic theories of Don Ihde concerning the auditory field. As Ihde notes, one of the strongest contrasts between sight and sound is that the visual field is frontal, in one dimension—we can see only what is in front of our eyes—whereas the auditory field is "*omnidirectional*."[5] Illustrating his point, Ihde notes:

> the visual field displays itself with a definite *forward oriented* directionality. It lies constantly before me, in front of me, and there it is fixed. As a field relative to my body it is *immobile* in relation to the position of my eyes, which "open" toward the World . . . [On the other hand] the auditory field as a *shape* does not appear so restricted to a forward orientation. As a *field-shape* I may hear all around me, or, as a field-shape sound *surrounds* me in my embodied positionality . . . I can switch my focal, auditory "ray" from one sound to another without even turning my head . . . My auditory field and my auditory focusing is not isomorphic with visual field and focus . . . [its] field-shape "exceeds" that of the field-shape of sight.[6]

Hearing individuals live in a constantly sound-saturated environment to which we are always subject and phenomenological practice has been foundational in studying sound cultures. But Ihde's principles also show in sharp detail how it is that a work such as Sow's should be understood as a sound art piece as much as a physical work of plastic sculpture. And it is its scale, studio process, and exhibition history that make it not only public, but social art.

## Jacques Coursil's *Trails of Tears*

In what follows I address a more obviously sonic work, trumpeter and composer Jacques Coursil's 2010 album, *Trails of Tears*, which was recorded in Paris and Fort-de-France, Martinique, and first released in France.[7] My discussion proffers a detailed examination of the album from musical, textual, and visual perspectives. Rather than making a generalized case for the album as a specific form, I investigate Coursil's project here as a case study in how taking a musical work as a mixed form may elucidate not only its content but political, aesthetic, and conceptual factors only accessible in an intermedia and networked reading such as those proposed in this volume. Like Sow's installation, *Trails of Tears* stages a parallel encounter between Black and Native American histories to listen for solidarity and mutual recognition. Like Sow, Coursil approaches the sounds of race thinking in the United States from the vantage point of an outsider, as a French citizen who sojourned there for a number of years as a professor and free jazz musician.[8] And, like Sow, Coursil points us toward alternative sensory accounts of colonialism and racialization that foreground the *sonic* as an essential counterpoint to the visual and textual. Intentionally embedding sound in this multimodal context, each work explores sound's capacity to realize (at least a desire for) solidarity. This chapter proposes to listen in stereo to these two very different sound objects—a traveling outdoor sculptural installation and an audio recording accompanied by a dense supplementary documentation. I have demonstrated this approach, "stereomodernism," more fully elsewhere, adopting an orientation toward soundings from different regions of the African diaspora that are in dialogue but not in perfect sync. What is key here is that this audiovisuality engages history on the grounds of contemporary expression.[9]

Stereomodernism works against nostalgia and the search for cultural retentions; rather, it explores how coeval articulations of modernity's quandaries generate sound texts with the potential to secure new sites and forms of solidarity. Solidarity is never a simple matter. With necessary circumspection, we consider these texts, marked by the most absolute history of race and racism in the post-Renaissance modern era, as encoding the desire for collective connections in the present may be realized in sonic terms. Furthermore, the critical tools of sound studies uncover how seemingly static forms of "surplus" noise can yield productive interpretations of the misrecognition, appropriation, and distortion that may attend such encounters. Listening to how US Native Americans and non-US Black encounters are sounded out in Sow's and Coursil's work, we can discern the aesthetic consequence of racialized history in the United States outside its national borders. Heard within earshot of each other, in the presence of all the languages of the world, as Martiniquan writer Edouard Glissant puts it, Black and indigenous histories come into relief as ineluctably bound up in the fashioning and maintenance of whiteness as not only a US-American national project, or even a European one but as a global affair.[10] That is the point at which world music becomes worlding music, and where Blackness becomes the portal through which to think—and indeed to hear—the Human most clearly.

*Trails of Tears* is a complex and multimodal object. Eschewing any pretense of citing indigenous musical motifs, Coursil's audio, visual, and textual discourse demands to be taken on its own terms, even as it engages all the sensory possibilities of the album as cultural form. In the track titles, A-B-A structure of the album, and liner notes, Coursil and his collaborators use a ternary form to hold a mirror between two acts of brutal forced displacement: the mass kidnappings of the Trans-Atlantic Slave Trade and the Trail of Tears that followed Andrew Jackson's Indian Removal Acts of the 1830s, decimating the Choctaw, Chickasaw, Creek, Seminole and Cherokee nations. These are not works *about* these two parallel atrocities. Instead, these are musical ruminations across time and space that *face* history. They train both the musicians' and their listeners' inner ears to recognize the monotony of crass evocations of "Red" Indians—whether noble savages or lurking threats at the borders of settlement—in everyday popular culture. Or even to notice the sheer absence of indigeneity in public consciousness. Arguably intellection has abandoned the ethical imperative of coming to grips with the ongoing consequences of slavery, dispossession, and generalized terror, rendering the call for reparations an appeal always already falling on deaf ears. Confronting this, Coursil's album requires that one *spend time* in the presence of the radically other. This intentional intimacy begins with each of the two ensembles playing in the presence of the other, then in the studio editing sounds into a shared time, and later as listeners partake in the recorded audio. The album is a relational device, intimately felt when each set of persons is impelled toward a collective act only possible in sonic terms: We must pay attention to both the opacity and the virtuosity of another collectivity, the original ensemble of indigenous persons expelled from their lands. To be sure, such attention falls short of what History requires of us. Nevertheless, the duration and durability of a collective auditory experience disrupts the patterns of dismissal that make US settler colonialism a strangely open secret and listening in the presence of these works produces an auditory Brechtian alienation effect. Multiple scenes of ensemble work demand that the listener dwell in the presence of those who are most radically other: the dead.

Coursil is a legendary figure in free jazz circles, both a musician and a scholar. Born in France to Martiniquan parents, he grew up hearing the cornet, a small instrument related to the trumpet, and Creole songs in his courtyard.[11] Between his last album of in the free jazz scene of the 1960s (*Way Ahead*, 1969) and his return album, *Minimal Brass* (2004), Coursil taught linguistics, philosophy, psychoanalysis, and Caribbean literature, and has written extensively about the active practice of listening in speech and music.[12]

Just as Sow asks us to attend to a material installation as a sound text, Coursil's use of the vanishing genre of the compact disc album exploits both its material and its sonic properties. Coursil's sound text is embedded in a matrix of accompanying materials that operate together to produce a musical discourse that appears to bear little indexical relation to the violent historical events named. The programmatic content of *Trails of Tears* emerges only via the notes by Bruno Guermonprez, producer, Bernard Vincent, historian, Edouard Glissant, a close friend and eloquent glosser of the trumpet's unique sonorities, and Coursil himself. This matrix of the liner notes, images, packaging, titles, and the like invites intermedial attention, and before turning to the visual and textual relays we come to the sonic experience.

Glissant declares in the liner notes that Coursil's music is that of a *lament*, "[n]ot a lamentation but its sparkling transcendence."[13] He hears a rhizomatic node linking the Native and Black condition in a new form of musical potentiality. If the Blues is less a sonic plaint and more an alchemy of trauma into trenchant truth, the *lamento* (echoing the French word for that most Caribbean of creatures, the manatee, or *lamantin*) similarly transform lamentation into something more than the mere sounding of inconsolable loss. Glissant's notes are significant not only for what he has to say but for what such an erudite voice suggests in terms of the philosophical engagement expected of the album's listener.

## A Mirrored Lament

As I have already alluded to, Coursil's collaborative *lamento* refuses to reduce a musical meditation on these human tragedies to a set of familiar but hackneyed tropes that have become sonic caricatures. There are no beating bass drums in 4/4 meter, no vocal ululations, no citations of melodies from Westerns. Nor are there any hints of timeless African authenticity, no reliance on polyrhythms, unaccompanied voices, call, and response. Coursil's music is abstract, variable, insistently modern, and by presenting three distinctive musical textures and styles of improvisation across the album, with three very different ensembles, it withholds any smoothing gestures or supposed coherence between sections.

The album was recorded over the course of one and a half years from May 2007 to January 2009, between Fort-de-France (Martinique), Montreuil (France), and Belleville (New Jersey), mapping a transoceanic transit across the Black Atlantic that the album's tracks also reference. The penultimate track is named "Gorée," after the island off the coast of Senegal from which so many Africans were deported and launched into the horrors of the Middle Passage, and in fact the premise of the pluralized *Trails of Tears* is the parallel passages of indigenous peoples across land and African captives across ocean—in other words this is a cartographic album. The track listing highlights the sections by including the location and date of each ensemble's recording session. The first section, on the Indian Removal, and the final section, on the Middle Passage, are recorded with the Martinique-based Cadences Libres. The second section, which addresses the Removal as a two-act tragedy, casts the Free Jazz Art ensemble for its first act, while a trio of Coursil, José Zébina on drums from Cadences, and Bobby Few on piano from Free Jazz in the second act. Coursil includes a further gloss on the history at hand:

I. *Nunna Daul Sunyi* (The trail where we wept) summarizes the ears of the conquered people who were deported, tears without water, dry-eyed and in silence.[14] Coursil chooses to begin with the Caddo name for the Trail of Tears, calling back the sound of indigeneity from its suppression within colonial languages and thus repopulating this territory with a Native American presence.
II. *Tagaloo, Georgia* is the name of a place lost forever in this uprooting with no return, the end of a world.
III. *Tahlequah, Oklahoma* is the end of a voyage for those who survived
IV. & V. *Free Jazz Art, the Removal Act I & II* relates what remains of the revolt against the greatest conquering enterprise in history, the colonial adventure, the Wars of the Whole World.
VI. & VII. [Gorée and Middle Passage] are the African response to the exodus of the Indians.[15]

The first three tracks, played by Free Cadences, are in diatonically related minor keys. The music begins in F minor, then flirts with the relative F major and with D minor, before closing in G minor. The minor confirms the *lamento* Glissant highlights in his liner notes. This opening set of tracks is highly consonant, with a breathy trumpet line above long chords that contribute a sense of harmonic stasis. The trumpet's repetitions of a stuttering melody, with each phrase played twice, and the melodic content of the phrases rendered tentative by the double-tonguing of the trumpeter.

A dramatic break in texture and harmonic language occurs in the longest track on the album, "The Removal *Act 1*." Here a much larger ensemble—including major figures of the 1960s free jazz scene Sunny Murray, Bobby Few, and Alan Silva— produces a fuller sound. The embrace of dissonance and unforeseen interactions generates a different quality of attention, while repetitions, section breaks, and instances of unison show that the work is tightly structured to organize the musical discourse of improvisation. Credits to J. Baillard for arranging all the tracks on the album offer further indication of the use of structure. This tension between form and free improvisation figures an almost agonistic relation between players, and, most crucially, the degree of risk involved in such a performance. The track closes with the sound of Jacques Coursil's enigmatic chuckle—a valve-release after the taut nerves of a single-take recording, perhaps, or a version of that essential Blues genius, the laugh that keeps from crying over the impossibility of adequately speaking (or sounding) historical trauma on such a scale. The interruption of musical tone with this most human of utterances breaks the sonic abstraction the track sustains for over eleven minutes reminds us that we are listening to technologically mediated audio signals. However, choosing not to turn to a sound engineer to edit out this moment reminds us that the artist is no slave to technology, but rather the radical presence of a virtuosic Black body. Laughter is involuntary, of little use value, excessive, but most essentially a uniquely human expression—it shakes the body, tears up the eyes, wipes out the breath. Coursil leaves it here, reminding us that it is not the instrument that sounds out, but the instrumentalist.

The smaller ensemble of "The Removal *Act 2*" (Coursil, Bobby Few from Free Jazz Art and José Zébina from Cadences Libres) creates an intersection between the two larger groups. While the instrumentation is thinner, this is the most sonically sublime track on the album, with Few's piano work producing sheets of sound with pedaling and block chords that traverse the entire keyboard to cloud and crowd the entire range of frequencies available to the acoustic space. This aural saturation is made all the more acute by the crescendos that sound a note of terror in each musical gesture.

Of all the tracks featuring Cadences Libres, "Middle Passage" is the most harmonically errant, wandering through A flat major, D flat minor, and chromatic progressions, and multiple plagal (IV-I) and perfect (V-I) cadences implying final arrival. That these gestures of resolution and arrival unfold in a track entitled "Middle Passage" suggests that the Black condition is one that arrives not on the shores, but in the hold of the dank slave ship.

The Cherokee nation is, of course, not virgin sonic territory in jazz. The well-known tune "Cherokee" started out as a rather saccharine mid-tempo song composed in 1938 by the British swing jazz artist Ray Noble. As an instrumental standard, "Cherokee" took on a life of its own, and throughout the 1940s and 1950s it was a staple in the brutally competitive jam sessions where the faster the band started the more likely young players trying to prove themselves were likely to embarrass themselves by failing to keep up. Even recordings captured this risk, the laying down of the gauntlet among musicians, as in Clifford Brown's version, on the 1955 album *Studies in Brown* recorded with the drummer Max Roach.[16] The recording opens with the trope of a four-beat ostinato tom-tom drum introduction with a heavy emphasis on the down beat. A listener immediately recognizes this as the kind of exoticized sound that classic Hollywood Westerns like *Stagecoach* (John Ford, 1939) trafficked in whenever signaling the entry of Native American characters, typically on the war. It is remarkable that in this burning version, even virtuosic drummer Max Roach struggles to keep tempo.[17] The conversion of this tune, and its implicit evocation of taking the Indian lover into one's arms—an act of possession, even if only metaphorical—remains an ironic undercurrent swirling below most important sites of power struggles on US and Indian territories, as in America's classical music, Jazz.[18]

Liner notes are a vanishing genre, a technology once essential to the meanings an album makes. *Trails of Tears* relies on this technology. On the cusp of disappearance, it is as copious as precarious. The notes comprise a booklet of sixteen pages, the first four in French, followed by a map spread across two pages, a set of photographs, four pages of English translations for the French, a page listing tracks, a page of credits identifying the performers, dates, and locations of recordings and, on the back cover, additional production credits, thanks and a dedication to Daniel Richard. The centerfold features each of the musicians' images and a map serving as a synecdoche for the material history of the Cherokee removal, moving in scale between the studio and the ethnoscape. The photographs signal that as a commodity the album represents the labor of nine individual recording artists who are the first, which is to say prime, sound collective in this project. The map, in stark contrast, cannot represent individuals but whole peoples who followed two itineraries, the land and water routes plotted over the faint outlines of states from Virginia, North Carolina, and South Carolina, on the eastern coast of the United States, to Kansas and Oklahoma in the mid-West. Significantly, the gray-scale outline of the states ends in open-ended squiggles rather than mapping the Eastern coast and Western state borders. Erasing coastal frontiers parallels the album's project of commensurating (literally, measuring together) the historical traumas of the Middle Passage and the Trail of Tears and specifically how they run into each other. If the interiors are the site of Native disaster, the coasts represent the annihilation of arrival for enslaved Africans. Where the borders of these coeval losses lie is left deliberately open in this visual representation—where one ends and the other begins is immaterial. These "trails of tears" use syntax to juxtapose the specificities of the historical removal of the Cherokee (and the other peoples the US government called the Five Civilized Tribes) to the unbounded suffering inaugurated by the age of discovery.

An excerpt from historian Bernard Vincent's *Le Sentier des larmes : le grand exil des Indiens Cherokees* briefly details the Trail of Tears.[19] A listener familiar with Black history mat notice that the Removals occur between 1830 and 1839, coinciding with the British abolition of slavery in 1834, and the treaties that eventually led to the French abolition of slavery in 1848. In other words, Coursil and Glissant, who both write regularly about an interarticulated global in Glissant's terms, *"le Tout-Monde,"* this historical context draws a silent parallel in the suffering of forced dispossession among Native Americans and enslaved Africans. The map and musical layering make this simultaneity audible. Glissant and Coursil have collaborated previously, as Coursil sets the poem "Les Grands Chaos" to music, reciting it on *Clameurs* in a voice reminiscent of Schoenberg's *Sprechstimme*.[20] *Trails of Tears*, by contrast, is entirely instrumental, but in a chiasmus of *Clameurs*, the poet sets music to text. One of the last texts Glissant produced before his death in 2011 proposes that Coursil remakes time in this performance, drawing attention to timbre and duration in this *lamento* so markedly different in mood from the fanfare and clamor of his two previous albums. Glissant observes

> the unexpectedness of silence, considered a space in which all sufferings and voices from afar are transformed; it is also the pursuit of an infinitely constituted echo . . . Coursil moves *time* aside from us, I mean to say, we learn with him to measure that time which comes to us from distant silences, from the scratched-out histories of peoples, and to put this beside us so that we may better appreciate it, and be able to *turn around* this time . . . the superb lament of this art is made of recomposed, returned silences, and of that interminable echo in which monotones and repetition sparkle like rocks of lava. This time which moves aside so slowly is the time of our awakening.[21]

Coursil, for his part, declares that "improvisation begins in this waiting, this suspension, which sometimes lasts and makes an unexpected silence . . . my body in as much as it breathes is the only soul I will ever have."[22] He adopts Glissant's perception of the trumpet as an echo, calling his work "un cri second," a second cry that returns and recites the tale of displacement in reverse. The page becomes a stage for the textual equivalent of an antiphon, two soundings answering each other. Indeed, echo is a concept Glissant's work has long addressed as an ethical recognition of irreducible difference translated into a sonic metaphor. Glissant's writing, too, embraces a technique far more familiar in (Afrodiasporic) music, repetition with a difference, or what he terms "*ressassement*," by republishing fragments of earlier works in subsequent volumes and by returning to specific characters and figures in his novels, theory, and poetry. Music in the jazz tradition, however free, is always connected to this theory of *ressassement*, both repetition and detour. After all, this is where the significance of earlier musical precedence is most clear, as in the speculative relation between a standard like "Cherokee" and Coursil's *Trails of Tears*. In this vein, Glissant's statements on opacity as a form of non-violent comprehension demands closer attention:

> [I] can conceive of the opacity of the other for me, without reproach for my opacity for him. To feel in solidarity with him or to build with him or to like what he does, it is not necessary for me to grasp him. It is not necessary to try to become the other (to become other) nor to "make" him in my image . . . Widespread consent to specific opacities is the most straightforward equivalent of nonbarbarism. We clamor for the right to opacity for everyone.[23]

## The Strain of Listening

The Trail of Tears is pluralized into *Trails of Tears* not as an act of commensurability or equivalency, but of recognizing a *Relation* in Glissant's sense of the term. In other words, Coursil sounds out a way to empathize with the suffering of the other without seizing the other's place, the other's understandings of that suffering, the other's specificity. A brief turn to etymology reminds us that the Greek origin, *empatheia*, represents a prepositional instance of feeling, to be in the pathos of the other. But its journey into English passes through the German coinage *Einfühlung*, where the /ein/ implies an entry into (*eingehen*, to enter, *einziehen*, to inhale). In other words, empathy entails a mode of entering into the other's feeling, without seizing it for one's self.

For Coursil the musical implications of Glissant's opacity have a literary precedent: "[P]oets have already sung of this."[24] And yet he turns to the prose of the Bajan poet, Edward Kamau Brathwaite, reminding us that a poet is often also thinker, historian, critic. Brathwaite's collection of essays, *History of the Voice* is grounded in both his cultural analysis of the Caribbean and in his training at Cambridge as a historian. Rather than Brathwaite's best-known statements on jazz in "Jazz and the West Indian Novel," Coursil cites the following: "Then came the destruction of the Amerindians which took place within thirty years of Columbus' 'discovery' (one million a year). It was necessary for the Europeans to import new labour into the area."[25] Those new laborers were, of course enslaved Africans, and thus Braithwaite makes the most explicit statement in the liner notes concerning not only a simultaneous but also causal relationship between the parallel histories of destruction wrought by settler colonialism and plantation economies over the *longue durée*.

In a more poetic register, Coursil chooses Glissant's preface to the section on The Trade (read the Slave Trade) in his long poem *Les Indes* (The Indies). The section begins: "It is a massacre here (in the reservoir of Africa), to compensate for the massacre over there."[26] Compensation is an odd, even discomfiting way to put this, but Coursil's training as linguist prompts a reading that goes beyond the surface of his words. To compensate, from the Latin *com-pensare*, where *pensare* is simply the habitual form of *pendere* (to weigh), is most often interpreted as to weigh against, but a more literal understanding points, rather, to weighing together. "Compensating," then, names the problem of weighing history and loss together, of bearing witness to the other's pain without occluding one's own, of bringing to bear one's own affective knowledge in order to enter *into* relation with the other. We hear, then, how Afrodiasporic and indigenous American subjects might empathize.

But what kind of labor is involved in empathizing? Who performs this work—musical, interpretive, affective? And to what extent can we claim this is a collective? These questions are all the more pressing given that the album's notes make scant mention of the liberation work, those acts of resistance that precipitated the Removal and of the civic work that prompted the backlash against Reconstruction. If the Cherokee leaders famously preferred misery and freedom to the imposition of US imperial laws with which they would have had to contend had they stayed in the territories that had been their homes for centuries, how might we reconceive of the refusal of work as rebellion on both slave plantation and Indian reservation? And what work of solidarity does this album suggest remains possible in a contemporary moment?

The liner notes leave unmentioned Coursil's ties to contemporaneous indigenous political struggles. Jason Weiss reports that in an interview with Coursil, he recalled "that *Trails of Tears* was conceived in the early 1970s when [Coursil] spent time among the Sioux in South Dakota and was told the story of the Trail of Tears." Coursil recalls

> I was there during the birth of the American Indian movement and I was deeply impressed by the seriousness of those people, who don't talk much. But when they say something, it's heavy. There are a lot of books about this, but that is nothing compared to people telling me things bit by bit. The musician always translates into music what they see and hear and smell and experience, so instead of making a theory out of it, I made music.[27]

Coursil's effort to honor those stories is chiefly reflected through the idiosyncratic *sound* of his trumpet, a fluttering, airy tone that respires in empathy with the belabored breathing of the long march, and the dark ship's hold. Just as walking through Sow's monumental installation memorializing the Battle of Little Bighorn demands that we listen to the cacophonies of contemporary traffic that cannot drown out history's plaints, thrust into the midst of this collective complicity in colonial amnesias both local and global, so too Coursil pulls us into his recovery of a possible way to affectively and effectively inhabit the polyphonic legacies of Black and Native dispossession. He draws us into posture of listening and keeps us occupied imagining the impenetrable challenge to decolonize our senses, rather than occupy land. We do not have to be settler colonialists. Sow and Coursil call us to participate in a sound collective of liberation, as pressing in the contemporary moment as in history's unresolved debts. This may be the only compensation we may offer, listening. Closely. In empathy.

## Notes

1. Tsitsi Jaji, *Africa in Stereo: Modernism, Music, and Pan-African Solidarity* (Oxford: Oxford University Press, 2014), 1–23.
2. Ousmane Huchard Sow, *Battle of Little Big Horn*, mixed media installation, 1999. For an art historical account of the installation including illustrations see Salah Hassan's essay "Native to Native: The Sculpture of Ousmane Sow." For images of Sow at work in his Dakar studio, see Magnum Photo's page "In the Studio: Ousmane Sow," by Martine Franck, https://www.magnumphotos.com/arts-culture/art/studio-ousmane-sow-martine-franck-dakar-fulani-masai-sudan/.
3. "Native American Ownership and Governance of Natural Resources." Department of the Interior, US government, https://revenuedata.doi.gov/how-revenue-works/native-american-ownership-governance/.
4. Huchard Sow, *Battle of Little Big Horn*.

5  Don Ihde, *Listening and Voice: Phenomenologies of Sound*, 2nd ed. (Albany: SUNY Press, 2007), 75.
6  Ibid.
7  Jacques Coursil, *Trails of Tears*, recorded in 2007–9. Sunnyside (2010), compact disc.
8  Coursil first came to United States in the 1970s, playing trumpet and teaching at the UN School before returning to France, and joining the faculty at Université de Caën Basse-Normandie. He returned to the United States to teach Caribbean literature in 2002, before retiring to Europe and dedicating his time to a second chapter of his recording career and to scholarly writing in the field of general linguistics.
9  Jaji, "Stereomodernism."
10  Édouard Glissant, "La langue qu'on écrit fréquente toutes les autres," *Le Monde des Livres*, March 16, 2006. See https://www.lemonde.fr/livres/article/2006/03/16/edouard-glissant-la-langue-qu-on-ecrit-frequente-toutes-les-autres_751247_3260.html.
11  Edouard Glissant, liner notes of *Trails of Tears*, Jacques Coursil, 11.
12  For more on Coursil's linguistics research see his written works, *La fonction muette du langage* (Point à Pitre: Ibis Rouge, 2000); "L'essence double de la musique: Pour une sémiotique de l'entente" (unpublished essay), and *Valeurs pures: Le paradigm sémiotique de Ferdinand de Saussure* (Limoges: Lambert-Lucas, 2015).
13  Edouard Glissant, "Jacques Coursil, and Time that Moves Aside," liner notes of *Trails of Tears*, 12. The original title in French reads "Jacques Coursil et le temps qui s'écarte," 4.
14  The use of the Caddo (Cherokee) language name for the Trail of Tears, Nunna Daul Sunyi, is an act of restituting language, one of those cultural resources stripped away by the Indian Schools.
15  Coursil's original French reads "sont les répliques africaines aux exodes indiens" (liner notes, 5). The French "répliques" captures the sense not only of responsive dialogue but also of systemic replication. Coursil, *Trails of Tears*, 12.
16  Clifford Brown–Max Roach Quintet, *Study in Brown*, recorded at Capitol Studio, New York City, February 25, 1955. The personnel included Clifford Brown on trumpet and Max Roach on drums. As the Leaders, and Harold Land (tenor sax), Richie Powers (piano), George Morrow (bass).
17  For more on this, see Claudia Gorbman's essay, "Scoring the Indian: Music in the Liberal Western," in *Western Music and Its Others: Difference, Representation, and Appropriation in Music*, ed. Georgina Born and David Hesmondhalgh (Berkeley: University of California Press, 2000), 238–9.
18  The last three lines of Cherokee's lyric reads: One day I'll hold you / In my arms fold you, / Cherokee (as recorded by Sarah Vaughn on *The Land of Hi-Fi*, 1956).
19  Bernard Vincent, *Le Sentier des larmes: le grand exil des Indiens Cherokees* (Paris: Flammarion, 2002).
20  Coursil, "Les Grands Chaos," track 6 on *Clameurs*, Universal, 2007, compact disc.
21  Glissant, liner notes of *Trails of Tears*, 11.
22  In the French original: "L'improvisation commence à cette attente, ce suspens, qui parfois dure et fait silence inattendu ... mon corps tant qu'il respire est la seule âme que je n'aurai jamais."
23  Edouard Glissant, *Poetics of Relation*, trans. Betsy Wing (Ann Arbor: University of Michigan Press, 1997), 190.
24  Jacques Coursil, liner notes of *Trails of Tears*, 12.

25  Kamau Brathwaite, *Roots* (London: New Beacon, 1984), 261. See also "Jazz and the West Indian Novel," *Bim* 11 (1967): 275–84.
26  Edouard Glissant, *The Indies*, trans. Dominique O'Neill (Paris: Éditions du GREF, 1992), 199.
27  Jason Weiss, "Jacques Coursil," *BOMB*, June 25, 2010, https://bombmagazine.org/articles/jacques-coursil/.

## Works Cited

Brathwaite, Kamau. "Jazz and the West Indian Novel." *Bim* 11 (1967): 275–84.
Brathwaite, Kamau. *Roots*. London: New Beacon, 1984.
Clifford Brown – Max Roach Quintet. *Study in Brown*. Recorded February 25, 1955. EmArcy, 1955, record.
Coursil, Jacques. *Clameurs*. Universal, 2007, compact disc.
Coursil, Jacques. *Jacques Coursil Unit - Way Ahead*. Recorded in 1969. BYG records, released 1969, LP.
Coursil, Jacques. *La fonction muette du langage*. Pointe à Pitre: Ibis Rouge, 2000.
Coursil, Jacques. "L'essence double de la musique: Pour une sémiotique de l'entente." unpublished essay, 2017.
Coursil, Jacques. *Minimal Brass*. Recorded in 2004. Tzadik Records, 2005, compact disc.
Coursil, Jacques. *Trails of Tears*. Recorded in 2007–9. Sunnyside, 2010, compact disc.
Coursil, Jacques. *Valeurs pures: Le paradigme sémiotique de Ferdinand de Saussure*. Limoges: Lambert-Lucas, 2015.
Ford, John, dir. *Stagecoach*. 1939; Los Angeles: United Artists. Film.
Franck, Martine. "In the Studio: Ousmane Sow." *Magnum Photo*. Accessed October 3, 2020. https://www.magnumphotos.com/arts-culture/art/studio-ousmane-sow-martine-franck-dakar-fulani-masai-sudan/.
Glissant, Edouard. *The Indies*. Translated by Dominique O'Neill. Paris: Éditions du GREF, 1992.
Glissant, Edouard. *Poetics of Relation*. Translated by Betsy Wing. Ann Arbor: University of Michigan Press, 1997.
Glissant, Edouard. "La langue qu'on écrit fréquente toutes les autres." *Le Monde des Livres*, March 16, 2006.
Glissant, Edouard. Liner notes of *Trails of Tears*, Jacques Coursil, Sunnyside, 2010, compact disc.
Gorbman, Claudia. "Scoring the Indian: Music in the Liberal Western." In *Western Music and Its Others: Difference, Representation, and Appropriation in Music*, edited by Georgina Born and David Hesmondhalgh, 234–53. Berkeley: University of California Press, 2000.
Hassan, Salah. "Native to Native: The Sculpture of Ousmane Sow." *African Arts* 32, no. 4 (1999): 36–49.
Ihde, Don. *Listening and Voice: Phenomenologies of Sound*. 2nd ed. Albany: SUNY Press, 2007.
Jaji, Tsitsi. *Africa in Stereo: Modernism, Music and Pan-African Solidarity*. Oxford: Oxford University Press, 2014.

Sow, Ousmane Huchard. *Battle of Little Big Horn*. Mixed media installation, 1999.
Vaughn, Sarah. "Cherokee." Recorded October 1955. Track 3 on *The Land of Hi-Fi*. EmArcy, 1956, record.
Vincent, Bernard. *Le Sentier des larmes: le grand exil des Indiens Cherokees*. Paris: Flammarion, 2002.
Weiss, Jason. "Jacques Coursil." *BOMB*, June 25, 2010. https://bombmagazine.org/articles/jacques-coursil/.

Part III

# Social Acoustics and Politics of Sound

# 7

# The Operator and the Final Girl
## Gender, Genre, and Black Sonic Labor in *The Call*

Allison Whitney

In the opening sequence of *The Call* (Brad Anderson, 2013), Los Angeles 911 operator Jordan Turner (Halle Berry) takes a call from a man reporting that burglars are breaking into his neighbor's house. As Jordan records the relevant details and conveys them to the police, the caller offers to handle the situation himself, saying, "I got a shotgun, want me to stop them?" Jordan immediately refuses the offer and strongly discourages him from such action, telling him to wait for the first responders. Jordan's management of this call encapsulates the role of the emergency operator: to assess a situation, to calm the person on the line, and to facilitate the action of police, all the while maintaining limits on the caller's agency, vetoing any proposed vigilante solutions and confining problem-solving to institutions of state power. In a later sequence, Jordan leads a group of trainees to speak with a newly appointed operator, Brooke (Jenna Lamia), and they ask which part of the job she finds most difficult. She replies that since she signs off the moment the police, ambulance, or fire department arrive, she struggles with "the not knowing, a lot of the time you don't know how it ends," and she finds this lack of closure emotionally taxing. Moments later, a call comes in from Casey (Abigail Breslin), a white teenage girl reporting her own abduction. Brooke is flustered by the lack of tangible information, as Casey is trapped in the trunk of a car, does not know her location, and is calling from a disposable phone with no GPS signal. Jordan steps in to handle the call, and thus begins a collaboration where Jordan will not only coach Casey on survival, but also use her attuned sonic literacy to discover the kidnapper's identity and location, and then surpass the boundaries of the operator's typical role to not only effect a rescue, but to exact vengeance. *The Call* intersects the genres of horror and crime thriller, specifically the procedural crime film that emphasizes the workings and failures of modern infrastructure. The film employs the spatial and sonic properties of the telephone system to further the narrative drive toward a collaboration of emphatically gendered and racialized genre-based character types: Casey as the horror film's "Final Girl," and Jordan, who merges the figure of the heroic Telephone Operator with the "Enduring Woman" trope of Black women in American horror. Their collaboration hinges on strategic use of sound technologies, their sensitivity to the acoustic properties of the spaces they occupy and listen to, and Jordan's vocal performances where she shifts tone and diction depending on the psychology of her interlocutor. The characters ultimately achieve the narrative resolution that Brooke and her fellow operators long for, and do so through a process where Jordan transcends her assigned role as a disembodied voice on the line, regaining her agency to appear in the flesh at the scene of the crime.

## Invisible Labor and the Disembodied Voice

In her essay "Weavers of Speech: Telephone Operators as Defiant Domestics in American Culture and Literature," April Middeljans addresses late-nineteenth and early-twentieth-century stories about the heroic exploits of operators, who were virtually all female. In these narratives, operators not only bravely stay at their posts to manage disasters, but they also transcend the parameters of their jobs and become sleuths, solving crimes through their observations of telephone communication: recognizing voices, overhearing conversations, and often violating company policy and police directives in order to save the day.[1] While automation has eliminated the need for telephone users to interact with operators on a regular basis, emergency calls still require human workers, and if one examines the film and television casting of such operators, they are overridingly portrayed by Black women. Indeed, in *The Call*, a significant proportion of Jordan's coworkers are women of color, and the film's first image of an operator is a close-up of a Black woman's mouth saying, "911, what is your emergency?" While this pattern reflects some real-world labor practices, it also speaks to a deeper symbolic logic based in anxieties about technology, embodiment, the ubiquity and fallibility of modern systems, and the simultaneous dependence on and disavowal of Black workers' technical and emotional labor.

The tendency to cast Black actresses in operator roles seems consistent with a broader relegation of such characters to "helper" status, where competent Black people aid the exploits of white protagonists. To offer one example, in an episode of *Law & Order: Special Victims Unit* entitled "911," where Detective Olivia Benson (Mariska Hargitay) tries to locate a kidnapped girl who calls for help on a cell phone, a series of characters who specialize in verbal and sonic communication, including several 911 operators, a language consultant who identifies the caller's accent, and an FBI agent who specializes in phone technology (Chandra Wilson), are all played by Black or biracial actresses. While it is not surprising that the casting practices of the media industry would tend to place Black performers in supporting roles, there is a more complex tradition here that touches on discourses on technology, labor, and the construction of racial othering. For example, Martin Kevorkian identifies a pattern in American film, television, and advertising, where experts in information technology are portrayed almost exclusively by Black actors. In *Color Monitors: The Black Face of Technology in America*, he explains how "race comes into sharp relief when computer use is depicted as difficult labor requiring special expertise. Time and again, in such scenarios, the helpful person of color is there to take the call—to provide technical support, to deal with the machines."[2] Not only do these characters exhibit technical skills, but they frequently possess augmented powers of perception—they are able to see, hear, or otherwise sense information in nearly superhuman ways. He cites numerous examples where Black figures, trained to their computer monitors, are able to make out subtle but essential details, from Cuba Gooding Jr.'s character in *Outbreak* (Wolfgang Petersen, 1995) noticing minute structural changes in a mutating virus, to Black characters in *The Hunt for Red October* (John McTiernan, 1991) and *Broken Arrow* (John Woo, 1996) locating otherwise undetectable military threats through their tracking systems.[3] While these portrayals valorize the characters' abilities, they also set up a troubling justification for an "imagined natural ethnic division of labor" that necessitates Black workers' remaining in prescribed roles where their abilities are most useful, and where their supporting white characters' heroism can seem merely a matter of practicality and wise use of human resources.[4] As we shall see, the factor of sensory acuity will be especially significant for characters in the role of telephone operator, where technical skill, sensitive hearing, and professionalized sonic literacy converge.

Kevorkian's analysis focuses on male figures as computer experts, but when it comes to the role of emergency operator, it is Black women who are ubiquitous. In her research on the history of workers in the Bell telephone system, *Race on the Line*, Venus Green explains how shifts in urban demographics in the 1960s led telephone companies to hire more Black workers, while retaining traditional gender divisions by having exclusively female operators, claiming that customers expected to hear a woman's voice when they called the company.[5] This premise was based in a long history of telephone companies cultivating a specific type of feminine image for their operators, entirely dependent on a fundamental fact of telephone communication, that of the disembodied voice. In *"Hello, Central?": Gender, technology, and culture*, Michèle Martin explains how in the nineteenth century, telephone companies began employing almost exclusively female operators, believing that "feminine" traits of physical dexterity and accuracy, along with gendered expectations for women to have moderate tempers and to be courteous and patient even with frustrated or irate customers, made women ideal for this kind of labor.[6] Further, operator training protocols were designed to homogenize these women's accents and diction, removing audible markers that might reveal their working-class or ethnic origins, and thereby cultivating an image of white and middle-class respectability.[7] Unlike the telegraph, where largely male operators would encode and decode messages but have little interaction with clients, early telephone users would speak with operators, and telephone companies perceived the benefit of conveying a "comfortable and genteel" image through their operators, an image that was specifically white and female.[8]

It is not unusual for industries to cultivate a public representation of their workers, and to use the gender, race, and class status of those figures to frame their corporate image. However, for telephone companies, assuring the public that the person on the end of the line met standards of respectability was especially important due to the technology's inherently disembodying properties. Being able to speak at a distance, in the absence of the interlocutor's body, is the purpose of the telephone, and for early users, this experience of the human voice estranged from its physical source could be disturbing, verging on the uncanny. Fictional works that used the telephone as a narrative device would often articulate this discomfort in ways that connected fear of disembodiment to a larger disruption of the social roles and hierarchies that seem to hinge on codes of bodily difference. As Tom Gunning explains in "Heard over the Phone: *The Lonely Villa* and the de Lorde Tradition of the Terrors of Technology," early cinematic telephone narratives included many frightening scenarios, including a particularly brutal "nightmare of masculine impotence" where a man, usually a husband and father, is forced to overhear a fatal assault on his wife and children while being too far away to intervene.[9] Telephone suspense scenarios might have tragic or heroic outcomes, often involving "rides to the rescue" initiated by a call for help, but in either case, the suspense relies on the interplay of particularly modern contingencies of time, distance, speed of vehicles, and fears about invading forces impinging on the domestic sphere and the women and children located there.

In American cinema, home invasion dramas consistently map out the domestic sphere as a space of embattled whiteness, often presenting the invading force as a racial or ethnic other, where the restoration of order depends on the physical return of white patriarchal authority to the home. At the same time, white domestic spaces and the broader social structures they represent are supported by the often-invisible activity of Black workers. Thus, the trope of the Black emergency operator in contemporary genres is consistent with American narratives where the salvation of the home depends on disembodied Black labor operating within systems of state surveillance and control. Indeed, in *The Call*, the opening montage shifts rapidly among 911 calls, each reporting a threatened family or domestic space: a sick child, family violence, a married couple experiencing a drug overdose, and, in Jordan's case, the caller who reports a burglary and wishes to effect his own "ride to the rescue." What is immediately clear is that these telephone workers respond to crises that quite literally hit close to home, and their effective intervention requires that they perform both emotional labor and management of state resources to maintain a status quo grounded in the domestic sphere.

In *The Call*, the central drama involves mobile phones and moving cars, and yet, the film's first violent sequence presents a more traditional telephone-based scenario that resonates strongly with the invasion tropes of American cinema and the ideologies they reproduce. A white teenage girl, Leah (Evie Thompson), calls 911 to report a man breaking into her house. Jordan takes the call and dispatches the police while advising Leah, who is calling on a landline with a cordless receiver, to hide in her bedroom, but not before opening a window to make it appear that she has escaped. Fooled by this ruse and unable to locate his target, the prowler begins to leave the house, just as the telephone is accidentally disconnected. Jordan reflexively calls the number back only to realize that the ring announces Leah's hiding place, allowing the intruder to find and assault her. Jordan is now in the position of the desperate figure in so many telephone-based narratives, forced to overhear Leah's cries for help while knowing that rescuers will not get there in time. The man picks up the receiver, and Jordan speaks to him, first deepening her voice to threaten him that the police will soon arrive, then shifting her tone to plead for Leah's life, carefully describing her as a "little girl" to inspire empathy. Alas, he simply signs off with the ominous statement "It's already done." Leah is later found murdered, and as we will discover, her attacker, Michael Foster (Michael Eklund), is the same man who will kidnap Casey.

Leah's call and its aftermath are traumatizing for Jordan, who feels responsible for her death. Even though her friends and coworkers reiterate that she is competent and that operators have limited agency, she continues to struggle with the incident, going so far as to keep a copy of Leah's obituary in her locker at work. Six months later she is no longer taking calls, but rather conducting trainings for future operators. As she leads trainees through the 911 center toward "the hive" full of "worker bee" operators, she emphasizes how crucial they are to surveillance of the city, stating that if "the hive goes down, the city goes dark." It is common for crime narratives to dramatize the systems that monitor urban spaces, but Jordan's emphasis on 911 as the "eyes and ears" of the city both introduces a bodily metaphor and hints at the broader implications of Brooke's concern about operators' persistent lack of closure. Here, it is useful to recall Kevorkian's point about how Black information workers are often described in terms of their acute sensory abilities. These workers function as the city's disembodied eyes and ears, but only until the arrival of first responders. They look and listen until the bodily presence of the police comes into play, and the prosthetic ears and voices of the operators are no longer needed. Jordan's guilt over Leah's death arises from a dissatisfaction with this lack of agency, and the closure it might offer.

This notion of embodiment has an important relationship to film sound, as Michel Chion argues in *The Voice in Cinema* with the concept of the *acousmêtre*. Chion explains how the cinema lends itself to an uncanny estrangement of the voice, where voices can appear on a soundtrack without being anchored in a visible body. Such voices are invested with a kind of omniscience and omnipotence, and films featuring acousmatic voices often have their eventual re-embodiment, and corresponding loss of their all-seeing power, as a narrative goal.[10] Meanwhile, part of Kevorkian's explanation for the semiotic alignment of Black workers with modern technological systems also draws upon anxieties about disembodiment, where "technology's capacity to disembody humanity, to take bodies out of the circuit of action, unconsciously projects technology onto the set of bodies it most fears".[11] I would argue that in this film, and in the figure of the operator more broadly, elements of both Chion and Kevorkian's commentaries on embodiment converge in the figure of the telephone worker, whose entire profession depends on disembodiment in both abstract and practical terms. Of course, while *The Call* hinges on these concerns, Jordan's voice is not acousmatic in the strictest sense, as from the perspective of the viewer, there is no question as to her identity or that the voice on the telephone is her own. However, from the distant perspective of her interlocutors, Leah, Casey, and Michael, she is a disembodied voice, and her estrangement from the physical space of the action is at the root of the film's power dynamics. As I will explain, Casey's rescue and the resolution of the narrative depend on a methodical reunification of sounds with the objects and subjects that produce them, and Jordan's conscious realignment of her body and voice.

## Final Girls and Enduring Women

*The Call* combines tropes of the crime thriller and horror genres, the latter being manifest both in the nature of Michael's crimes, where he mutilates his victims in rituals that rehearse an obsession with his deceased sister, and in the character of Casey, who develops into the archetype of the female survivor, the "Final Girl." In her 1987 essay "Her Body, Himself: Gender in the Slasher Film" Carol Clover theorized the Final Girl through an examination of horror films of the 1970s and 1980s. Her analysis demonstrated how horror, so often denigrated for its seemingly blatant and simplistic misogyny, could offer remarkably complex engagements with gender politics. The Final Girl possesses a level of agency and subjectivity that is rare in mainstream cinema, and her standard characteristics challenge traditional femininity: not only does she often have a gender-ambiguous name, but she is intelligent and independent, mechanically competent, vigilant to danger, and while she is attractive, she remains virginal, or at least chaste in comparison with her more sexually active friends. Clover notes how these traditionally "masculine" traits equip such characters for heroic deeds, and while she cautions against any facile equation of female gumption with progressive politics, she notes how Final Girls provide a rare point of cross-gender identification for male audiences. More broadly, the direct evocation of and challenges to gender roles allow for stories and scenarios that grapple with social norms on both literal and figurative levels, demonstrating how "gender is less a wall than a permeable membrane."[12] In the decades since Clover coined the term, the concept of the Final Girl has entered popular discourse, and filmmakers have quite consciously adapted the figure to engage with gender politics and with the genre itself, as evinced in such film titles as *The Final Girls* (Todd Strauss-Schulson, 2015) and *Final Girl* (Tyler Shields, 2015) where Abigail Breslin plays the title role, two years after portraying Casey in *The Call*.

When we meet Casey, she is spending the day at the mall with her friend, Autumn (Ella Rae Peck). Autumn reveals that she has a second cell phone, a gift from her boyfriend Raul, allowing them to communicate without her parents' knowledge. Casey immediately identifies this second phone as evidence of an unhealthy and controlling relationship, saying, "he bought that to keep tabs on you," and rejecting Autumn's romantic perception of his gesture. As their conversation continues, Autumn offers to help Casey find a boyfriend, but Casey is not interested. After Autumn realizes that she is late to pick up her brother, she hurriedly departs, leaving behind Raul's cell phone. Casey notices the phone and places it in her back pocket. On a practical level, this sequence allows Casey to acquire the phone she will soon use to call 911, but it also performs a function typical of the horror film, of using a conversation among female friends to designate a lead character as a Final Girl.[13] After she is kidnapped, Casey also comes to endure the typical Final Girl trials—observing the killer's violence against others (stabbing and burning male bystanders who try to intervene), discovering the remains of other victims, trying to flee the killer's traps, receiving physical scars in the process, and ultimately vanquishing the killer by appropriating his weapons, strategies, and powers.[14]

While Casey is firmly in the Final Girl role, *The Call* is distinct from a conventional horror film in that she is not the protagonist. Rather, this is Jordan's story, and our first impressions of Jordan are of course in her role as an emergency operator in a procedural crime drama. And yet, Jordan shares many of the Final Girl's characteristics, in that her name is not clearly gendered, and she possesses intelligence, independence, and technical skill. While these attributes would seem to neatly correspond to her ensuing mentorship of Casey as the narrative tropes shift toward horror, the racial dynamics of the film, and of the genres it draws upon, complicate that equation. In *Horror Noire: Blacks in American Horror Films from the 1890s to the Present*, Robin Means Coleman discusses how there are Black female survivors in horror, but with the important distinction that "Final Girls tend to be White."[15] She coins the term "Enduring Woman" for the Black female horror protagonist, and she "endures" because the monsters she escapes or defeats represent vast systematic oppression, so even if the initial threat is neutralized, there is no "finality" to her struggle.[16] While Enduring Women are powerful and resourceful, an important distinction is that they are less masculinized, and often assertive in their sexuality. In Jordan's case, we observe her flirtation and physical intimacy with her police officer boyfriend Paul (Morris Chestnut) early in the film, and she is clearly comfortable in her adult sexuality, in contrast with Casey's adolescent reluctance. And yet, when Jordan invites him to spend the night, her plan is interrupted by Leah's call and the ensuing emotional trauma. When Paul goes to Jordan's apartment later that evening to check on her, she does not let him in, and while he comforts her after the discovery of Leah's body, assuring her that "I'll be here when you need me," months go by before they speak again.

While Jordan's truncated romantic life might seem incidental to the narrative, the film's clear evocation of genre conventions and character types, including Final Girl, Enduring Woman, heroic operator, and Black information technology worker, particularly one with special sensory skills, points to an intentional hybridity, and one that is consistent with Halle Berry's star persona. As Charles Burnetts explains in "Re/Inventing Halle Berry: Mixed-Race Stardom and the Melodrama of Female Victimhood," Berry's career has required constant negotiation of concepts of "artificiality and hybridity" due to her multifaceted professional background as a beauty pageant contestant, fashion model, Oscar-winning actress, and also her status as a biracial woman.[17] While "Berry's mixed lineage allows her to straddle the boundaries of racial differentiation and thereby maximize her casting potential and economic feasibility as a star," it also lends itself to roles that map genre hybridity onto her characters' race and gender.[18] Not only does Berry's star persona convey this concept of hybridity, but Jordan uses the telephone to bridge these generic roles, first by actively coaching Casey on what amount to Final Girl survival strategies, drawing upon her own as-yet-latent Enduring Woman and telling her, "I want you to fight. Can you fight for me, Casey?" Second, she employs her sensory acuity and sonic literacy to discern auditory clues overlooked by the police, and, finally, she embraces the defiant operator trajectory, leaving her post to become a bodily presence in the film's final confrontation, where she and Casey collaborate to entrap and neutralize the villain.

## Acoustics, Space, and Power

The sequence at the mall concludes with Casey's abduction, setting in motion her Final Girl narrative. As she walks through the parking lot, Michael startles her with his car, then drugs her into unconsciousness and forces her into the trunk, making sure to destroy the smartphone she had just used to speak with her mother. When Casey wakes up, the car is on the highway, and she soon hears Autumn's phone ringing in her back pocket, indicating a voice mail from Raul. She uses that phone to call 911, whereupon Brooke hands the call to Jordan. Not only does Jordan offer encouragement to Casey, but the details of this sequence make clear that Jordan's capacity to effect a rescue, and the broader generic and symbolic significance of their collaboration, depend on the acoustic properties of the spaces the characters occupy, and their capacity to control and interpret sounds. When Casey wakes up, the first thing she sees is light entering the trunk through a stereo speaker, while she hears a muffled rhythmic sound. As she gradually emerges from her delirium, she begins to claw at the speaker, looking for a way out. The next shot is a close-up of that same speaker from the car's interior, before another cut to a close-up of Michael's ear. On a practical level, Michael is playing loud music in order to obscure noises from his abductee, and he turns up the volume as Casey begins to kick and scream. Meanwhile, there is a marked change in audio quality as we shift between the trunk and the car's interior, and, while that is consistent with a level of realism, it also makes a point of differentiating spaces by their acoustic properties. As Casey will soon realize, her survival will depend on her strategically discerning and controlling the sounds around her.

The stereo is playing Taco's 1982 rendition of the Irving Berlin song "Puttin' on the Ritz," and while Michael's taste in music might seem merely eccentric, the song has a complex history when it comes to race in American cinema.[19] In *Puttin' on the Ritz* (Edward Sloman, 1930), the eponymous song appears in an ensemble musical number using an interracial cast—a first in Hollywood history. While the Black and white dance troupes do not mix on stage, they do appear simultaneously, with the Black dancers on an upper platform and the white ones below. Berlin's original lyrics describe the practice of white New Yorkers visiting Black nightclubs in Harlem, and the intersection of class and racial mixing in their enjoyment of Black fashion, music, and culture. Subsequent revisions of the lyrics obscured these racial references, but over fifty years later, Taco's version was a matter of controversy as the music video featured performers in blackface. Of all the songs Michael might listen to, it is striking that he chooses one so clearly tied to points of contact, however strained, among Black and white populations, as well as the appropriation of Black culture and the objectification of Black people.

While Michael's music plays, the interaction between Casey and Jordan, itself a joint effort of Black and white characters, sets into motion a process of subversion. Much in the way that Casey reverses the disempowering agenda behind Autumn's cell phone by using it to help herself, "Puttin' on the Ritz" gives Casey the sonic cover she needs to converse with Jordan without being discovered. Jordan counsels Casey in strategies to identify the car, such as listening to the road to determine that she is on the freeway, breaking through a tail light so she can stick her arm out to attract attention, and pouring cans of paint she finds in the trunk onto the road to leave a trail. These strategies are somewhat successful, catching the notice of bystanders and giving the police clues to narrow down her location. For example, while stopped at an intersection, another motorist, Alan (Michael Imperioli), notices the paint and signals to Michael that something is wrong with his car. Fearing that he will be detected, Michael drives away in a panic, further arousing Alan's suspicions and inspiring him to follow Michael to an isolated parking lot. Michael opens the trunk to assault Casey and prevent her from jeopardizing his plan, only to be interrupted by Alan. In a turn that is typical of Final Girl narratives, this potential male rescuer ultimately fails, and Michael steals his car and beats and murders him in front of Casey. It is during this confrontation that Michael nearly discovers Casey's phone as Jordan's voice calls out from Casey's back pocket. Casey begins screaming to cover the sound, and Jordan realizes what is happening just in time to mute the audio, recalling her critical mistake where she had inadvertently announced Leah's location to her killer. What becomes clear is that while Jordan is a very skilled operator, the parameters of her job require her to be acoustically present as a listener while remaining otherwise imperceptible at the scene of the crime, audible only to her interlocutor.

As Michael reaches the entrance to the underground lair where he intends to murder Casey, he finally discovers the phone she has been using to speak with Jordan. He holds it up to his ear, and just as in Leah's case, Jordan uses a shift in tone and diction to persuade him that surrender is his best option. By now, the police have managed to collect enough clues to identify Michael, and Jordan tells him as much, assuming an authoritative tone, uttering his full name, informing him that police have interviewed his wife and children, and that his best course of action is to turn himself in. As before, while her voice conveys authority, she cannot act directly on the scene, but can only warn of the imminent arrival of the police. Michael is unmoved, so she shifts to a more pleading tone, but just as in Leah's case, he concludes the call, saying, "It's already done." Jordan's recollection of this phrase makes her realize that this is in fact the same person who killed Leah, which both underlines the futility of her disembodiment and builds her resolve to rescue Casey.

Casey wakes up, now tied to a chair, and Michael begins to wash, cut, and style her hair as a prelude to her murder. In the midst of this process he puts on a cassette recording of Culture Club's 1983 song "Karma Chameleon."[20] By now we have come to understand that Michael's rituals are an attempt to return to his childhood when his sister was still alive, and this would seem to explain his taste for 1980s pop music. And yet, just as with Taco's "Puttin on the Ritz," he again chooses a song whose music video contains an explicitly racialized narrative. While the lyrics of "Karma Chameleon" evoke deception and destructive relationships, the original music video presents a fantasy scenario set in "Mississippi – 1870," where a cast of Black and white performers whose costumes suggest equal social status cavort on a steamboat named *The Chameleon*.[21] They collaborate to foil a thief, a white man who steals jewelry and cheats at cards, throwing him overboard once they discover his crimes, thereby restoring the boat to an idealized space of racial harmony. As the song plays, Casey makes her first attempt at an escape, freeing her hand and spraying hairspray into Michael's eyes before running out of the room. As is common in horror films, she takes several wrong turns in Michael's disorienting labyrinth, but the size and dimensions of the space are strongly denoted by the echo of the music. While Michael recaptures Casey just as she makes the gruesome discovery of a bloodstained bedroom, it is conspicuous that at yet another juncture in Casey's enactment of the Final Girl persona, the diegetic music connects to a narrative of interracial collaboration, in this case an alternate-history version of a desegregated American South where Black and white characters work together to foil an exploitative man. In the sequences featuring "Puttin' on the Ritz" and "Karma Chameleon," not only are the songs thematically significant but the music also emphasizes the acoustic properties of the vehicle and the bunker, allowing Jordan and Casey to then use these acoustic cues strategically to provide sound cover for their collaboration, identify and navigate spaces, and eventually neutralize danger. While it is not unusual for sound designers to denote the size and shape of spaces with acoustic cues, the conspicuous combination of these specific songs' racial connotations with the sonic mapping that ultimately aids the heroines' collaboration and victory allows the music to frame Jordan and Casey as a utopian "post-racial" collective marked by song.

Dissatisfied with the progress of the investigation, Jordan, not unlike the heroic operators of nineteenth-century fiction, begins to take matters into her own hands, and this process involves both a targeted use of her particular skills as a sound worker and a gradual move away from her official duties and the systems of state power they serve to coordinate. She begins by laboring outside her working hours, analyzing the recordings of Casey's call for clues. In a scene that recalls *Blow Out* (Brian DePalma, 1981) she discerns and isolates a mysterious clanking sound. Meanwhile, the police have determined that Michael owns an isolated property, but their inspection of that house offers no actionable clues. Jordan decides to investigate the house herself, and in her search, she discovers photographic evidence of Michael's incestuous obsession with his sister, a girl with long blonde hair, just like the girls he targets in his murders. As she is leaving the house, Jordan stops to listen and recognizes the same clanking sound from the recording, leading her to a flagpole standing directly over Michael's bunker. It is important to note that the reason she had the flagpole recording in the first place is due to Casey's collaboration in carefully concealing the phone, and also Jordan's skill as an operator, keeping Michael on the line for an extended period as she argued for Casey's release while buying time to trace the call. While he destroys the phone before they can finalize the trace, he stood beneath the flagpole long enough to reveal this crucial spatial clue. Thus, Jordan's combination of emotional labor and sonic literacy allows her to collect and analyze evidence and arrive at an acoustic mapping of the scene that will lead her to Casey.

Having located the trap door to the bunker, Jordan's first impulse is to call 911, but her phone slips from her fingers into the pit below. She climbs down to retrieve it, only to discover that in the underground space, there is no cell signal. In the typical logic of the horror film, when a character discovers that there is no cell service, it is a sign of imminent danger, as they have lost their capacity to call for help, or to provide vital information to other characters. However, in this case, Jordan's rescue mission requires that she detaches herself from the telephone system and its mechanisms of disembodiment. Up until this point in the narrative, her efficacy as an operator had depended on her acoustic presence and physical absence, but in the final stage of the rescue she steps out of her assigned position and discovers her own power as an Enduring Woman. Jordan hears distant cries, and instead of climbing back up and completing her call, she follows the sound—indeed, virtually all the cues she uses to navigate the space are sonic—and she ultimately makes the same gruesome discovery of the bedroom from the "Karma Chameleon" scene. Jordan hears Michael coming and hides behind a wardrobe door, stifling her gasps of horror as he plays out a grotesque ritual involving Leah's dismembered scalp, while the camera reveals the bloodstained walls and furniture. In many respects, this scene rehearses the scenario of Leah's call, with Jordan as an observer to violence, both unable to intervene and acutely aware that survival depends on silencing her own voice. Once Michael leaves the room, Jordan's audible gasps return to the soundtrack, underlining how her status as observer relies on both her invisibility and her inaudibility. However, in the scenes that follow, she will not only rescue Casey and avenge Leah, but do so in a manner that emphatically re-embodies that voice and her agency.

Jordan follows Michael back to the room where he intends to mutilate Casey, sneaks up behind him, and knocks him unconscious. Casey asks, "who are you?" but Jordan does not answer, instead repeating reassuring phrases as she cuts through the tape on Casey's wrists. After a moment, Casey recognizes her voice, exclaiming "Jordan?!" Michael comes to and tackles Jordan, forcing her head into a sink of water, in an attack that is explicitly silencing in its approach, at which point Casey counter-attacks, slashing him with the scissors he had used to cut her hair. These reciprocal actions are emblematic of Jordan and Casey's collaboration, but they are also consistent with horror genre conventions where Final Girls and Enduring Women typically turn the villain's weapons against them.[22] A chase ensues, and they climb out of the bunker with Michael close behind, the fight choreography making it quite explicit that Jordan and Casey's escape is a collaborative effort, exhibiting a remarkable degree of coordination. As Jordan starts to run away, Michael turns to chase her, but Casey, still holding the scissors, stabs him in the back. Then, as he turns to retaliate, Jordan rushes him and shoves him back into the hole, giving Casey the opportunity to kick him in the face to knock him out.

Jordan pulls out her phone to dial 911, but pauses when Casey shouts "Wait!" She and Jordan glance down at Michael's prone body, and then at one another. Casey smiles and cocks her head, whereupon they formulate a plan that will not only punish Michael, but also craft a false narrative employing both the common-sense premises of law enforcement and traditional genre conventions to effect vigilante justice and bypass institutions of state power. Moments later, they are back underground, but this time they have chained Michael to the chair where he had previously immobilized his victims, recalling Casey's awakening in the "Karma Chameleon" scene. Jordan says, "You should have listened to me, Michael Foster. You could have turned yourself in when you had the chance." He stares at her for a moment, then recognizes her voice and says, "Oh, you're the operator." He laughs briefly and continues "I thought you'd be taller ... so when do the police get here?" Jordan replies with a sarcastic tone, "the police?" then turns to Casey, who continues in a lilting voice: "I escaped. Jordan found me in the woods. And you ... disappeared." They turn to leave, and realizing that they intend to abandon him to die in his own trap, Michael shouts after Jordan, "you're just an operator, you can't do this!" She replies with Michael's stock phrase, "It's already done," before she slams the door, ending the film.

Michael's emotional responses in this final scene are quite revealing. His initial reaction to being tied up is one of panic and anger, but once he recognizes Jordan as "the operator" he becomes strangely relaxed, as he believes that the authorities will soon arrive, and his fate will follow a predetermined path within the justice system. Even though the presence of an emergency operator at a crime scene is unprecedented, and suggests that the rules may have been rewritten, in this moment he still perceives her as incapable of acting directly on the scene. His attitude is reflected in his laughter, and his "I thought you'd be taller" comment—a cliché phrase used to dismiss one's adversary, and specifically to discount her physical capabilities, which surely include her race and gender. Even though Jordan and Casey have literally overpowered him, he is still unable to see past the perceived limitations of her profession. Indeed, it is fair to say that he never really "thought" about her height, or any physical attribute at all, except in a framework where Black and female labor remains disembodied and invisible. Even in the face of his now grim reality, he still has confidence in the power structures and the predictable outcomes that the system provides. He is therefore all the more shocked when Jordan and Casey decide to, in effect, write a new ending to the story.

To appreciate the implications of Jordan and Casey's plan, it is significant that they do not kill Michael directly. Rather, they abandon him to a fate that is clear, in that he will surely die of dehydration or starvation, but also, at least for him, terrifyingly open-ended. When Jordan says, "It's already done," she reclaims a phrase intended to underline the powerlessness of the telephone interlocutor. Michael took a sadistic pleasure in saying it, for even though his victim was still alive, he knew that temporal and spatial distance would preclude any obstacles to his violent plans. While Casey had appropriated Michael's hair scissors in a move that rebukes his fetishization of her physical traits, Jordan's seizing his catchphrase is especially fitting for a character whose power is tied to her voice. Jordan repurposes his spoken phrase to represent her own satisfaction at attaining the narrative and emotional closure so often denied of operators, while also placing Michael in the position of the helpless listener. In addition to the vengeful satisfaction of this conclusion, the film again deploys its merging of genre conventions to underline Jordan's accomplishment and Casey's collaboration. When they relate the false narrative of Casey's escape and Jordan's fortuitously finding her in the woods, they fabricate a conventional Final Girl narrative, where a white teenage girl rescues herself, only to be assisted by a Black supporting character who brings her to safety. Not only does their story conceal Michael's location and prevent his rescue by authorities, but it also shields Jordan from any consequences of stepping out of her operator role. Jordan and Casey know that systems of state power will not afford them satisfaction or closure, so they instead co-author an alternate history, one that is in fact closer to the victories of the Enduring Woman, whose capacity to save the day, and save herself, may not dismantle the broader social structures that support the villain's violence, but do hint at the possibilities of a visionary new reality.

## Conclusion

One of the points Kevorkian makes about the figure of the Black technology worker is that their explicit immobility, where they must remain at a work station, facilitates the mobility of white protagonists while neutralizing fears about the agency of people of color.[23] With this dynamic in mind, it follows that Michael imagines himself as the protagonist in his own story, capable of moving through urban spaces undetected, at least until Casey and Jordan begin their collaboration. In vanquishing him, they render Michael immobile and hidden, and then use his inability to take action as an opportunity to author not one but two narrative resolutions. In her efforts to defeat him, Jordan comes to reject the immobility of the operator's station and embrace her capacities as an Enduring Woman, all the while imparting Final Girl strategies to Casey. Jordan and Casey flip the script on Michael, and instead of fulfilling the expected roles of women in crime dramas by calling in male authorities to rescue them, they devise their own conclusion to the story and mete out their own punishment. Not only do they render Michael "already done," but they also concoct a fictional narrative of escape and rescue, one that leaves Michael's fate unresolved. Jordan's initially acousmatic voice provided Casey with the emotional and tactical support she needed to embrace her Final Girl, while Casey, newly embodying that role, supports Jordan in reintegrating her body and voice, and transcending systems of state power to save the day on her own terms.

## Notes

1 April Middeljans, "'Weavers of Speech': Telephone Operators as Defiant Domestics in American Culture and Literature," *Journal of Modern Literature* 33, no. 3 (2010): 44–6.
2 Martin Kevorkian, *Color Monitors: The Black Face of Technology in America* (Ithaca, NY: Cornell University Press, 2006), 2.
3 Ibid., 23–4.
4 Ibid., 24.
5 Venus Green, *Race on the Line: Gender, Labor, & Technology in the Bell System, 1880-1980* (Durham, NC: Duke University Press, 2001), 221.
6 Michèle Martin, *"Hello, Central?": Gender, Technology, and Culture in the Formation of Telephone Systems* (Montreal: McGill-Queen's University Press, 1991), 59.
7 Brenda Maddox, "Women and the Switchboard," in *The Social Impact of the Telephone*, ed. Ithiel de Sola Pool (Cambridge, MA: MIT Press, 1977), 268.

8. Kenneth Lipartito, "When Women Were Switches: Technology, Work, and Gender in the Telephone Industry, 1890-1920," *The American Historical Review* 99, no. 4 (1994): 1084.
9. Tom Gunning, "Heard over the Phone: The Lonely Villa and the de Lorde Tradition of Terrors of Technology," *Screen* 32, no. 2 (1991): 191.
10. Michel Chion, *The Voice in Cinema*, trans. and ed. Claudia Gorbman (New York: Columbia University Press, 1999), 28.
11. Kevorkian, *Color Monitors*, 2.
12. Carol J. Clover, "Her Body, Himself: Gender in the Slasher Film," *Representations* 20, no. 20 (1987): 208.
13. Ibid., 204.
14. Ibid.
15. Robin R. Means Coleman, *Horror Noire: Blacks in American Horror Films from the 1890s to Present* (New York: Routledge, 2011), 131.
16. Ibid., 132.
17. Charles Burnetts, "Re/Inventing Halle Berry: Mixed-Race Stardom and the Melodrama of Female Victimhood," in *Star Bodies and the Erotics of Suffering*, ed. Rebecca Bell-Metereau and Colleen Glenn (Detroit, MI: Wayne State University Press, 2015), 311.
18. Ibid., 313.
19. Taco, "Puttin' on the Ritz," recorded 1982, track 3 on *After Eight*, RCA Records, audio cassette.
20. Culture Club, "Karma Chameleon," released October 1983, track 1 on *Colour By Numbers*, CBS Studios, audio cassette.
21. Culture Club, "Karma Chameleon – Official Music Video," directed by Peter Sinclair, released 1983, 2005 Digital Remaster, video, 4:00, https://www.youtube.com/watch?v=JmcA9LIIXWw.
22. Clover, "Her Body, Himself," 202.
23. Kevorkian, *Color Monitors*, 4.

# Works Cited

Anderson, Brad, dir. *The Call*. 2013; Stamford, CT: WWE Studios. Film.

Burnetts, Charles. "Re/Inventing Halle Berry: Mixed-Race Stardom and the Melodrama of Female Victimhood." In *Star Bodies and the Erotics of Suffering*, edited by Rebecca Bell-Metereau and Colleen Glenn, 309–26. Detroit, MI: Wayne State University Press, 2015.

Chion, Michel. *The Voice in Cinema*. Translated and edited by Claudia Gorbman. New York: Columbia University Press, 1999.

Clover, Carol J. "Her Body, Himself: Gender in the Slasher Film." *Representations* 20, no. 20 (1987): 187–228.

Culture Club. "Karma Chameleon." Released October 1983. Track 1 on *Colour By Numbers*. CBS Studios, audio cassette.

Culture Club. "Karma Chameleon – Official Music Video." Directed by Peter Sinclair. Released 1983. 2005 Digital Remaster. Video, 4:00. https://www.youtube.com/watch?v=JmcA9LIIXWw.

DePalma, Brian, dir. *Blow Out*. 1981; New York: The Criterion Collection, 2011. DVD.

Green, Venus. *Race on the Line: Gender, Labor, & Technology in the Bell System, 1880–1980*. Durham, NC: Duke University Press, 2001.

Gunning, Tom. "Heard over the Phone: The Lonely Villa and the de Lorde Tradition of Terrors of Technology." *Screen* 32, no. 2 (1991): 184–96.

Kevorkian, Martin. *Color Monitors: The Black Face of Technology in America*. Ithaca, NY: Cornell University Press, 2006.

Lipartito, Kenneth. "When Women Were Switches: Technology, Work, and Gender in the Telephone Industry, 1890–1920." *The American Historical Review* 99, no. 4 (1994): 1075–111.

Maddox, Brenda. "Women and the Switchboard." In *The Social Impact of the Telephone*, edited by Ithiel de Sola Pool, 262–80. Cambridge, MA: MIT Press, 1977.

Martin, Michèle. *"Hello, Central?": Gender, Technology, and Culture in the Formation of Telephone Systems*. Montreal: McGill-Queen's University Press, 1991.

McTiernan, John, dir. *The Hunt for Red October*. 1991; Hollywood: Paramount Pictures. Film.

Means Coleman, Robin R. *Horror Noire: Blacks in American Horror Films from the 1890s to Present*. New York: Routledge, 2011.

Middeljans, April. "'Weavers of Speech': Telephone Operators as Defiant Domestics in American Culture and Literature." *Journal of Modern Literature* 33, no. 3 (2010): 38–63.

Petersen, Wolfgang, dir. *Outbreak*. 1995; Los Angeles: Warner Brothers. Film.

Shields, Tyler, dir. *Final Girl*. 2015; Barnaby: Nasser Entertainment. Film.

Sloman, Edward, dir. *Puttin' on the Ritz*. 1930; Los Angeles: United Artists. Film.

Strauss-Schulson, Todd. *The Final Girls*. 2015; Culver City: Sony Pictures Home Entertainment. BluRay.

Taco. "Puttin' on the Ritz." Recorded 1982. Track 3 on *After Eight*. RCA Records, audio cassette.

Wolf, Dick, and Patrick Harbinson, writers. *Law & Order: Special Victims Unit*. Season 3, episode 7, "911." Directed by Ted Kotcheff. Aired October 4, 2005. NBC Universal Television. Universal Pictures Home Entertainment, 2008. DVD.

Woo, John, dir. *Broken Arrow*. 1996; Los Angeles: Twentieth Century Fox. Film.

8

# Bohemian Like You

## The Construction of Cool Sound Collectives in Serial Television

Florian Groß

In 2001, the telecommunications company Vodafone released the television commercial "How Are You?"[1] The one-minute-long spot consists of a fast-cut montage of several scenes that show a succession of young, energetic people that are connected to the world through mobile communication: a woman at a festival, people in urban night life, friends in a limousine, a group of teenagers spontaneously diving off a cliff into the ocean, a father switching his office job for a visit to the zoo with his son, teenage boys on motor scooters gawking at an attractive woman, a Bohemian party in the countryside, a rescue mission for offshore surfers, and crowds watching a Manchester United home game. In all of these instances, the instant connectivity of (then relatively new) mobile communication seamlessly blends into the active, nonconforming lifestyle of these people—even in a traffic jam, it helps the car passengers to connect with the world outside the gridlock. All of this is accompanied by and expressed through "Bohemian Like You" by Portland-based alternative act The Dandy Warhols.[2] This upbeat rock song extradiegetically supports the commercial's content and emotive message, both through its music and through lyrics that run down a couple of stock elements of modern Bohemia, such as coolness, temp jobs, playing music in bands, or eating vegan food. Notably, both song and montage are interrupted by a scene at the opera, where a ringing cell phone disrupts a ballet performance of *Swan Lake*,[3] annoying the audience and embarrassing the phone's owner before the clip cuts back to the rock song.

What the song provided for the spot (and brand) was therefore a lively, young image—and subcultural cachet, through its creator as well as sound. As such, it might be considered yet another example of a corporation branding its product as "young" and "hip" through the use of alternative (music) culture. This idea of corporate culture co-opting alternative culture for commercial reasons has often been used to explain its attraction for these companies. To a certain extent, this approach certainly applies, as the commercial potential of young, alternative culture for the branding and selling of goods has become an ever-increasing factor in contemporary consumer capitalism.[4] However, the question arises whether this process of co-optation has to be seen in necessarily antagonistic terms. Do we really deal here with capitalist business interests exploiting authentic alternative culture for commercial reasons alone, or is it maybe more accurate to consider them in much more interrelated and co-dependent terms? Read like this, the clip's juxtaposition of alternative and opera music might be more telling concerning the function of alternative culture in this context than a pure focus on questions of co-optation. From the perspective I want to propose in the following, alternative culture defines itself against the established mainstream of cultural spheres such as the opera just as much as it symbolically distances itself from commercialism and mass culture, but it does so concurrently with a late capitalist culture that also increasingly works this way.

Analyzing the interaction between alternative culture and capitalist media culture along these lines allows us to also find accurate answers to the question why alternative music has not only become part of television commercials, but rather a staple element of (serial) television in general. During the last three decades, licensed popular music has evolved into an increasingly conventionalized and commercialized element of television series, and a genre that has been particularly prominent in this regard is the vague entity known as independent or alternative pop and rock. While the idea of co-optation is never far from approaches to this topic, I want to claim that a study of this sonic element of television also benefits from an approach that accounts for the dynamic relationship between alternative "cool" culture and established mainstream culture in the context of commercial popular culture. The endlessly elusive term "cool" is both central to my argument and notoriously difficult term to define. For the purposes of this chapter, I will start from Pountain and Robins' "rough working definition" of cool as "an oppositional attitude adopted by individuals or small groups to express defiance to authority . . . which does not announce itself in strident slogans but conceals its rebellion behind a mask of ironic impassivity."[5] In line with this understanding of cool, this chapter seeks to outline how the series' non-diegetic soundtracks as well as their diegetic representation of music cultures feed into and represent a cultural position whose emphasis on subcultural, nonconformist distinction from the mainstream has become central in contemporary Western capitalism and postindustrial middle-class culture.

Through an analysis of the series' general use of alternative rock in their soundtracks as well as close readings of selected scenes, this chapter shows how independent music performs a dual function in serial television: First, its inclusion positions and collectivizes the audience(s) as ideal recipients of this form of culture; second, the producers' conspicuous consumption of subculturally inflected music positions both their serial texts and themselves within this cultural paradigm as well. However, rather than reading this exclusively in terms of commercial mass culture co-opting "authentic" subcultures, I read the complex interaction between alternative rock and serial television as a reciprocal aesthetic process and negotiation *within* contemporary capitalism along the lines of "cool," "hip," and "creativity." Within television culture, it furthermore adds another dimension to contemporary discourses on "Quality TV." Shows that prominently feature alternative music and culture do more than merely creating and exhibiting a certain kind of improved serial television; in addition, they emphasize the normative layer of alternative stances on mainstream culture as yet another element of heightened quality.

In the following, I approach the social and cultural repercussions of musical representations on television. For this, I want to focus on the way how instances and representations of alternative music in selected television shows negotiate a liminal stance between supposedly antagonistic cultural fields that has become a popular subject position in contemporary consumer culture. As one representative element of a vastly more complex sociocultural development, these negotiations stand metonymically for the complexities of indie culture vis-à-vis commercial imperatives. In concrete form, they exemplify the way how alternative music acts always traverse the fine line between staying "indie" and "selling out," that is, retaining the oppositional, autonomous cultural stance they supposedly have and sacrificing this separation from mainstream society and culture for the opportunity to make a profit.[6] Following Aslinger's claim that "[t]elevision criticism must wrestle with the industrial norms and cultural connotations of licensed music to more fully understand how licensed tracks mobilize meaning,"[7] I want to approach how alternative music performs this function of creating meaning in a sense that goes beyond specific instances of characterization or narrative and rather enhances and enriches the meaning of a given text as a whole. Moreover, I want to outline how the cultural politics of Quality TV-series and their collective (aural) address rely to a considerable part on a cultural positioning that is associated with elements of subcultural spheres—and therefore hope to come up with more precise answers concerning their cultural and commercial appeal.

## Cool, Creativity, Capitalism

Since the 1990s, several studies have made the claim that a particular combination of artistic innovation and cultural subversion has become a potent driver of a mode within consumer capitalism that thrives on difference, individuality, and nonconformity. Here, especially Thomas Frank's book *The Conquest of Cool* (1997) and its concept of "hip consumerism" are crucial for the argument I want to make.[8] In this study, Frank identifies developments in postwar US business and consumer culture that have created "a cultural perpetual motion machine in which disgust with the falseness, shoddiness, and everyday oppressions of consumer society could be enlisted to drive the ever-accelerating wheels of consumption."[9] Therefore, connected to the counterculture of the sixties, the cultural critique of mass society and consumerism spilled over into the broader culture and laid the foundation for ever-new forms of consumption that are defined as a consumer-critical stance, yet still revolve around the act of consumption. Frank's crucial take on the matter, however, is to show how this was not (only) a cynical business scheme created by Madison Avenue to trick unwitting consumers in ever-more purchases, but rather a sentiment that also resonated *within* American business culture. In his study, Frank shows how—similar to the way the Beats and later the counterculture rebelled against conformist postwar America—young advertising executives rebelled against the ossified hierarchies and stifling conventions of corporate America in their own capitalist field.[10]

Frank's study focuses on developments in the advertising and menswear industries of the 1960s, but he identifies a resurgence of this kind of consumption since the 1990s.[11] One of his primary examples in this matter is the music industry. In the article "Alternative to What?" Frank polemically outlines how the music industry discovered subcultural forms of guitar rock during the Grunge explosion and Second Wave of Punk Rock and used them as viable commercial vehicles. For Frank, "[t]here are few spectacles corporate America enjoys more than a good counterculture, complete with hairdos of defiance, dark complaints about the stifling 'mainstream,' and expensive accessories of all kinds . . . New soundtracks, new product design, new stars, new ads. 'Alternative,' they call it. Out with the old, in with the new."[12] While certainly less nuanced than Frank's overall approach, this sneering quote nevertheless captures the central dynamic between alternative and mainstream cultures, a dynamic whose dialectic is at once more complex and less antagonistic than any binary juxtaposition may make it seem. What Frank describes as an elaborate history of the complex process behind what is commonly associated with terms such as "selling out" or "co-optation" has subsequently become one of most central aspects of twenty-first-century consumer capitalism: the increasing popularity, especially in so-called "creative industries" and the subject and consumer positions related to this, to define oneself in contrast to "the mainstream."[13]

Frank's approaches to US consumer culture have proven to be very influential—either directly or indirectly—for more recent critical accounts of alternative culture and coolness in the context of consumer capitalism. For instance, in *The Rebel Sell* (2005), Joseph Heath and Andrew Potter outline how various multinational companies and brands use this dynamic to engender increased consumption in otherwise saturated markets.[14] Next to more specific analyses of consumer culture, this approach can also be found in recent sociological studies of creativity, which are particularly rich in this regard. Starting with the rather celebratory coinage of a "creative class" by urbanist Richard Florida, and next to studies such as Jim McGuigan's *Cool Capitalism* (2009), several studies have appeared that outline how the development identified by Frank, Heath and Potter, and McGuigan has become a widespread phenomenon that increasingly defines Western capitalism in general.[15] According to these studies, aspiring to a nonconforming, irreverent, individualistic, and at the same time culturally productive subject position is as structurally relevant for producers of goods, services, and ideas as it is for consumers. However, it has become something that is not only aspired to *in* contemporary culture, but even facilitated, rewarded, and demanded *by* advanced capitalist societies. In his study *The Invention of Creativity*, German sociologist Andreas Reckwitz has defined the central element of this development as the "creativity dispositif."[16] According to Reckwitz, subjects living and working in times of "aesthetic capitalism" are increasingly oriented toward the production and experience of new cultural forms and practices. These new forms of culture are marked by highly aestheticized signs of newness, difference, and defiance, a development that has made this kind of cool creativity become a Foucauldian dispositif in its own right.[17] "Being creative" in the sense of being individualistically, autonomously culturally productive is not only a subjective and collective desire, but also a social and cultural imperative—for everyone. In Reckwitz's words: "In late modern times, creativity embraces a duality of the wish to be creative and the imperative to be creative, subjective desire and social expectations. We want to be creative and we ought to be creative."[18] Simply—and only slightly hyperbolically—put: From the point of view suggested by Reckwitz, it has become inconceivable for people living in Western late capitalist societies to *not* want to be creative. I want to use this pervasive understanding of the role of "cool" and creative culture in contemporary society to analyze how alternative music is used across a variety of recent television shows.

## The Alternative Soundtrack of Serial Television

From singer-songwriter to punk and from indie pop to neo folk, guitar-oriented music with subcultural appeal has become a prominent element of the televisual soundscapes since the 1990s, as even a cursory glance shows: Death Cab for Cutie and other indie acts are not only featured as extradiegetic music, but also physically enter the stage within the diegesis of *The O.C.* (Fox, 2003–7),[19] a character listens to punk rock band Hot Water Music on *One Tree Hill* (The WB/The CW, 2003–12),[20] The Dandy Warhols provide the theme song for *Veronica Mars* (UPN/The CW, 2004–7, Hulu 2019),[21] a song by indie folk group The Decemberists appears on the soundtrack of period drama *Mad Men* (AMC, 2007–15),[22] and *Californication* (Showtime, 2007–14) even features the title of an album by the alternative rock superstars Red Hot Chili Peppers in its very name.[23] Particularly the teen drama *The O.C.* was associated with the promotion of a particular kind of music, and it is through a closer look at this show that I want to develop possible explanations for and functions of alternative music on recent serial television. The use of sound and music in television is, of course, highly diverse and ranges from the creation of certain moods, character development, and narrative functions to merchandising/cross-promotion and, for the lack of a better term, cultural politics.

In this regard, *The O.C.* is a fitting example due to a number of reasons. First, it features copious amounts of licensed music and primarily relies on 1990s and early 2000 alternative rock by bands such as Phantom Planet, Death Cab for Cutie, Nada Surf, Rooney, and many more. Next to its extradiegetic use of alternative rock music, it also features actual performances by bands from this genre. While the show's first season only includes a single concert performance by Rooney,[24] the show's second season introduces the location of the Bait Shop, a concert venue on the Newport Pier, where several prominent bands appear (e.g., The Killers, Modest Mouse, The Walkmen, Death Cab for Cutie, The Subways, Rachel Yamagata). Given its heavy use of alternative music, it is no surprise that the series also served as a vehicle for the promotion of new releases by bands such as the Beastie Boys, who debuted their single "Ch-Check It Out" in the first season episode "The Strip."[25] Out of this prominent and consistent use of popular music, no less than six official soundtrack compilations were released between 2004 and 2006.[26]

Second, through its "music supervisor" Alexandra Patsavas, it featured a prominent creative figure that went on to work in a number of further television series and films (most notably among them: *Mad Men*, *Grey's Anatomy*, *Twilight*, *Gossip Girl*). There, she continued to implement her particular—and increasingly recognizable—choice of music and thus extended this influence well over and beyond the show itself.[27] Beyond this matter of influence, though, another aspect is the liminal status of *The O.C.*; while its status as a self-consciously aware, at times metafictional prime-time show with high production values and A-list personnel such as actor Peter Gallagher, director Doug Liman, and producer McG makes it compatible with contemporary notions of Quality TV, the fact that it belongs to the rather denigrated genre of the teen drama and was broadcast by the Fox network makes it a more ambivalent candidate in this regard. As such, its use of alternative music highlights and intensifies cultural debates on the selling out of a supposedly anti-commercial music scene—after all, if you are featured on something so obviously commercialized as a teen drama series on a broadcast network, people will irrevocably doubt your alternative status. Emily Zemler has described this as "*The O.C.* effect," when a band such as Death Cab for Cutie appears in

> Fox's *The O.C.*, a show whose characters may be a bit trashy but whose writers seem to have impressive musical taste. This sudden upsurge of music that is— gasp!—considered to be good in the eyes of elitist hipsters and rock critics alike is surprising, proving extremely beneficial to bands like the Shins and Death Cab for Cutie in terms of exposure—and cold hard cash. It seems like a winning formula for all parties involved, except for the original fans, who smell sell out.[28]

Somewhere in-between gaining exposure, increasing record sales, selling out, and relinquishing creative control to television executives who use alternative music for their serial texts, Zemler argues, lie the potential pitfalls and benefits for the bands themselves. But what about the other side, that is, the shows themselves?

In its own negotiation of this debate on alternative music in commercial culture, the show follows a similar path. For instance, in the show's second episode, three characters discuss their musical tastes. Ryan Atwood (Ben McKenzie), a kid from the wrong side of the tracks, has just arrived in wealthy Newport Beach, while the nerdy outsider Seth Cohen (Adam Brody) and the popular girl Marissa Cooper (Mischa Barton) may have been living next door to each other since birth, but the latter has so far been hardly aware of the former's existence. On a drive back home, the following conversation unfolds:

> Ryan: "What kind of music do you listen to?"
> Marissa: "Right now, punk."
> Seth: "Yeah, I am sorry, but Avril Lavigne doesn't count as punk."
> Marissa: "Oh yeah? Well, what about The Cramps? Stiff Little Fingers? The Clash? Sex Pistols?"
> Seth: "I listen to the same music as Marissa Cooper? I think I have to kill myself."[29]

This dialogue tells us a lot about the points I want to make here. First, music emerges as an important marker of adolescent/young adult identity. Second, through the character of Seth Cohen, *The O.C.* expresses a cultural binary in which it is highly improbable that the beautiful and supposedly shallow Marissa listens to something as culturally profound as punk rock. If at all, supposedly commercialized and watered down "sell out" versions of this "authentic," traditionally independent and alternative kind of popular music are conceivable as the music of choice for her. Then, however, comes the surprise for both Seth and the audience, both of which were ignorant so far that Marissa could be any more than the popular girl living next door. United by the series' activation of cultural dichotomies along the lines of mainstream and alternative and the belief that someone like Marissa would never listen to punk or other forms of cool alternative culture, both learn in this scene that she is aware of and refers to classic idols of the genre.[30] This, however, does not create an affirmative reaction, but rather utter dismay on the side of Seth, who obviously fears that the possessive—and to a certain extent very gendered—exclusivity of his musical taste is threatened. In this regard, the choice of Avril Lavigne instead of, say, Good Charlotte or Blink-182 is telling. All of these artists stand for a popular version of 1990s punk that has often been accused of being hardly more than a cheap sell-out, yet the one referred to by Seth to disparage a girl's musical taste is conspicuously female. Therefore, with curiously gendered undertones, this scene emphasizes a larger point the series is at pains to drive home from the start: Seth Cohen is the sole true avatar of alternative culture. As the audience had already learned from the posters in his bedroom, Seth is a fan of alternative and independent music, with posters of bands such as Alkaline Trio or Ramones hanging on his walls. This, however, is celebrated at the same time that it is questioned and negotiated by scenes such as the one just quoted. Next to the fact that Seth is depicted as a band shirt–wearing, comic book–collecting, and skateboard-riding character, this scene emphasizes the kind of cultural dichotomy that the series both represents and perpetuates at the same time that it problematizes its central tenet of mutual exclusivity.

Ironically enough, though, while the show's soundtrack does lean heavily toward guitar-oriented music, actual punk rock as such plays only a minimal role here. Rather, it is a specific kind of singer-songwriter and indie music that is virtually omnipresent in the series. Ben Aslinger uses the term "sonic fingerprint" to describe the specific aural style of a given series, and *The O.C.*'s sonic fingerprint is certainly identifiable as "indie."[31] In the show, the prominent, and also heavily commodified, use of this kind of popular music is juxtaposed with a deprecatory representation of ostentatious wealth and rampant consumerism, therefore notably distancing itself from two basic aspects of the contemporary teen drama that have been fundamental elements since its inception through *Beverly Hills, 90210* (Fox, 1990–2000).[32] In *The O.C.*, the gated communities of Newport Beach, California, lavish mansions, and standardized "McMansions" are inhabited by financial traders, lawyers, and real estate brokers, and the children of these people attend a private high school—but the show's resident alternative character Seth Cohen openly despises them. Significantly enough, next to the gendered elements outlined earlier, this is also related to matters of class. As an inhabitant of Newport Beach, Seth Cohen comes from a privileged background and therefore has the opportunity to despise the crass materialism he is surrounded by—and which he can take for granted. Ryan, on the other hand, is introduced as the "kid from the wrong side of the tracks." While his reaction to upper-class living ranges from amused puzzlement to fisticuffs, he is shown to be much less antagonistic to the life it can provide for those living it. This difference is even emphasized by expressions of musical taste—while Seth listens to "cool" 2000s alternative rock, lower-class Ryan is introduced as being a fan of "uncool" 1980s hard rock and bands such as Journey. This binary opposition, however, is once more problematized during the course of the series: In the first-season episodes "The Escape," Seth and his eventual girlfriend Summer fight over the music playing on the car radio and Seth closes the argument by shouting, "Do not insult Death Cab!"[33] Half a season later, during another car ride and in a curious reiteration of this scene that had established Seth's musical taste in very vocal terms, Ryan forbids other car passengers to make fun of his musical taste in a very similar fashion: "Do not insult Journey!"[34] Therefore, while Ryan's class-bound musical taste might still be lacking in relation to the series' sonic fingerprint, his self-aware echoing of Seth's declaration exhibits a degree of ironic detachment that makes the difference between the two much less pronounced than they are initially made out to be. Fittingly, over the course of four seasons, and in close relation to Ryan outgrowing both his sociocultural background and original musical taste, he increasingly becomes a more ironic, self-aware—in short, cool—character.

In effect, with *The O.C.*, we are dealing with a lavish soap opera that includes its own detractors in its narrative, and sound plays an instrumental role in the series' act of distancing. In the words of Faye Woods:

> The programme draws on elements of *Dynasty*'s prime-time soap camp and excess in its depiction of Newport society, yet works to contrasts this world with its protagonists. Presented as 'authentic' outsiders, Seth and Ryan view this society through a self-aware, slightly mocking distance. However, the emotive portions of *The O.C.*'s soundscape – plaintive indie-rock, new folk or singer-songwriter – allow music to provide the emotional connectivity that its playfully ironic pose could potentially deny. As a result, the programme allows a dual level of engagement for its audience, who could revel in the 'knowing' soap opera plotting and the boys' witty banter, yet at the same time be drawn to the emotional realism of its musical moments.[35]

With respect to the relation between televisual music and serial narration, Faye Woods uses the example of *The O.C.* to illustrate how popular music can function both as an identity marker and a storytelling tool. Its indie soundtrack in connection to the show's representation and negotiation of alternative culture serves a dual function here; on the one hand, it allows for producers and audience alike to create a cultural niche beyond the supposed mainstream of the culture it is created in as well as the televisual genre it originated from. While this is hardly ever an unambiguous endeavor that is discursively problematized *and* commercially exploited by the series at the same time, it nevertheless forms the backbone of the series in terms of identification—at least for all those viewers who expect something different from a teen drama than the endless display of the romantic adventures and conspicuous consumption of supernaturally beautiful characters. Equally ambivalent is the storytelling function served by alternative music in *The O.C.*: On the one hand, its signification of difference is crucial to the show's attempt to distinguish itself from "standard" teen fare and thus create novelty in an established genre. On the other hand, the emotional, often enough outright sentimental quality of indie music does indeed do more than merely providing a contrapuntal indie soundtrack to its mainstream romantic plots; in several instances, the extradiegetic songs *amplify* the series' melodramatic effect.

Woods also highlights how the show's recurrent use of popular songs as leitmotifs within individual episodes create "a playfully reflexive relationship between narrative, soundscape, and audience."[36] Across episodes, Woods argues that Jeff Buckley's version of "Hallelujah" (originally written and performed by Leonard Cohen)[37] introduces a "series leitmotif" that relies on repetitive variation within the series: "Its use draws on the track's cultural connotations of melancholic indie credibility, yet also accumulates resonance within the narrative, developing musical meaning beyond individual moments and episodes to span the programme as a narrative whole."[38] I want to connect the two aspects juxtaposed here by Woods, namely, the "indie credibility" and its accumulating effect across serial texts. While the emphasis on narrative is less pronounced here, the recurrent effect of the use of popular music that is culturally connotated as indie or alternative has considerable effects and repercussions in the series as such. One important element consists of the way how the show creates the possibility to appeal to audiences that would normally not watch teen dramas and reward this kind of collective through the repeated inclusion of music that matches their tastes. Another, related element is the serial creation of newness, which in my opinion works in a very specific way here and further explains the attraction of alternative music to television series. In "Alternative to What?" Thomas Frank claims that "[t]he culture industry is drawn to 'alternative' by the more general promise of finding the eternal new, of tapping the very source of the fuel that powers the great machine."[39] Read along these lines, the use of alternative music provides the series with a "fresh" take on an established genre by self-reflexively critiquing basic elements of the genre through supposedly antagonistic cultural stances and tastes. This is relevant insofar as serial texts, both within their narratives, but even more importantly in relation to other, preceding or simultaneously released texts, always seek out novelty, but in a highly repetitive formal structure. More precisely, as critics such as Kelleter or Mayer have shown, seriality is marked by a constant process of varying repetition, and one possibility of variation in repetition is the association of serial texts and genres with different cultural fields.[40]

To approach this, I want to draw on another aspect that has become increasingly important in television studies since the 1990s. Associated with prime cable channels such as HBO and Showtime and an increasing amount of critically acclaimed network dramas, the term Quality TV has become virtually synonymous with a normative increase in production value and critical acclaim that supposedly differentiates this kind of serial television during the "post-network era" from previous instances of TV.[41] On the surface, alternative and independent cultures do not play a role in the context of Quality TV, at least not in the specific sense outlined earlier. However, the central artistic negotiations that can be identified with respect to this discourse share common ground and—in their emphasis on transgressive artistic expression in close relation with commercial success—are especially fitting with respect to Reckwitz's creativity dispositif. While countless studies on Quality TV have appeared since Thompson's initial study,[42] his basic list of characteristics still works well as a condensed summary of his most central ideas. In effect, his claim that Quality TV is "not regular TV" juxtaposes mass culture and more authentic forms of popular culture in a manner similar to alternative culture's rejection of the mainstream.[43] Moreover, Thompson's emphasis on the "noble struggle against profit-mongering networks and non-appreciative audiences" by the creative personnel that produces this kind of TV echoes creative culture's image of mainstream culture.[44]

Some critics have read Quality TV in relation to Bourdieu's theory of distinction as the televisual version of the expression of upper and lower class through the associated taste formations along the lines of "high" and "low."[45] However, in relation to the cultural theory outlined earlier, I want to suggest a slightly different focus. From the perspective I want to propose here, many Quality TV-series represent a different, less vertically stratified, and less economically deterministic form of distinction. Similar to the Vodafone commercial, and in close relation to cool culture's wary stance on elite and respectable society,[46] these series distance themselves as much from uniform mass culture as they do from traditional forms of high culture. The culprit here is not consumption per se, but rather forms of consumption that do not give the individual enough space for individual expression—and one of the prime venues through which this is expressed is music, both as an element to create and express this kind of distinction and as a cultural phenomenon through which the series can self-reflexively negotiate their own complicity. While this has become a prominent (upper-)middle-class imaginary, its identity with class positions is less pronounced than in orthodox Bourdieuean approaches. At the same time, its congruence with the alternative/mainstream divide is striking and reaches far beyond a single televisual text.

## In Lieu of a Conclusion: An Outlook

With *The O.C.*, I have focused on a case study that belongs to the genre of the teen drama. However, I want to argue that this phenomenon goes well beyond this specific televisual genre. Even though the negotiation of alternative music on TV certainly plays a crucial role in teen dramas and may be the televisual origin for the function I want to stress here, it is by now only one of many instances in this regard. Therefore, by way of a conclusion, I will briefly introduce two other examples that illustrate how these instances are evident throughout the—by now increasingly defunct—TV schedule.

One such example is the prime cable dramedy *Californication*. In this show about Hank Moody (David Duchovny), a struggling and at the same time conspicuously cool author in Los Angeles, alternative culture is an important part of the series' representation of (its own) creativity. When Hank's agent and best friend Charlie Runkle (Evan Handler) tries to revive the languishing career of Rick Springfield (played by himself), he has one thing in mind: "First, we get you some indie-cred. Let's find you your *Wrestler*, like Mickey Rourke."[47] Runkle's rather professional seeking of "indie-cred" for an obviously uncool artist is telling here insofar as *Californication* does very much the same and does so in a significantly open fashion. Overall, similar to *The O.C.*, and already evident from the Red Hot Chili Peppers allusion in the show's title, alternative (music) culture plays an important role in the series' construction of creative culture on TV. Moreover, many of Hank's books are named after albums by thrash metal icons Slayer (*God Hates Us All*, *South of Heaven*), and the episode that features a lengthy flashback detailing how Hank met his on-off girlfriend Karen (Natascha McElhone) and how she became pregnant is called "In Utero,"[48] after Nirvana's final album. The morning that Karen finds out that she is expecting a child, she learns of Kurt Cobain's suicide while we listen to the album' lead single "Heart-Shaped Box."[49]

Through elements like these, the high-profile television show *Californication* uses alternative culture in its serial text to both signify difference from the mainstream and broaden its appeal at the same time that it lays bare its motivation for doing so. Newman has shown how alternative and indie cultures are both instrumental in the construction of "an authentic, autonomous alternative to mainstream media" and frequently complicit in its commercial appeal.[50] Similar to *The O.C.*, *Californication* makes use of this effect to symbolically distinguish the series, its creative hero(es) and its creative team from the mainstream. In the case of *Californication*, the representation of contemporary alternative culture goes hand in hand with frequent allusions and visual cues that evoke the counterculture of the 1960s and 1970s. The fact that *Californication*'s creator Tom Kapinos used to be on the writing staff of *Dawson's Creek* (The WB, 1998–2003) may be yet another indicator that this form of youth culture has its televisual roots in the teen drama genre but has become as much a part of general serial television as youth culture in general has become a central element and trajectory of contemporary capitalist culture.

Another example is the period drama *Mad Men*. This show about the advertising industry in the 1960s was often hailed for its historical accuracy and meticulous reconstruction of American life in this decade. Significantly enough, one crucial element of the series is its depiction of the transformation that took place in US business and consumer culture; essentially, the series can be read as an indirect representation/adaptation of Thomas Frank's study.[51] Next to historical events, social mores, and set design, the show's celebrated historical authenticity also included its soundtrack, which primarily consists of extradiegetic period pieces. However, sometimes the show deviated from this by anachronistically including song selections from its own historical production context. The first instance of this was the Cardigans' 1997 alternative pop song "Great Divide"[52] in the show's second episode "Ladies Room."[53] The instance I am primarily interested in, however, occurs at the beginning of the second-season episode "Maidenform."[54] In this episode's opening montage, the song "The Infanta"[55] by alternative folk band The Decemberists accompanies a three-part scene in which we see three of the show's female protagonists getting dressed in front of mirrors. Introducing an episode that follows these women's different gender-related trajectories in the 1960s, the song's upbeat tempo signifies a feeling of departure, while the song's lyrics about the celebrated arrival of an infant princess echo the gender-related theme of the episode. As such, it vaguely reflects the episode's theme as much as the Cardigans' song did in the show's second episode, in which the audience witness an early instance of the "Great Divide" between the series' protagonist Don Draper and his wife.

While these aspects arguably play an important role in the show's usage of (modern) songs, I want to stress another element. To put it in Woods' terms: Next to a storytelling tool, music functions here as an identity marker as well. In this regard, I think that it is significant that in Charlie Wells' account of *Mad Men*'s musical anachronisms, he opens the section on "The Infanta" as follows: "Hipsters loved this emphatic juxtaposition when it opened a mid-second-season episode on white-versus-black, Jackie-versus-Marilyn brassieres."[56] While he continues to dive into the possible textual and sonic resonance between the song and the scene/episode, he does not go into further detail why it appealed so much to "hipsters." For this, a little bit of context on the band might be necessary. The Decemberists were founded in 2000 in Portland, Oregon, and are known for their 1960s folk revival–inflected version of alternative rock. The band's musical aesthetics are characterized by opulent arrangements and whimsical lyrics, and they have become one of the staples of contemporary independent music in the United States. Taken together with the fact that Portland—where not only The Decemberists, but also The Dandy Warhols, originated—is one of the nation's hotbeds of alternative culture, the connotations of the selection become even more apparent.

This reference to contemporary alternative culture, however, once more gains a more ambiguous quality if juxtaposed with the series' diegetic representation of 1960s counterculture. While *Mad Men* is often read as an indictment of the proverbial "Man in the Gray Flannel Suit," its depiction of Beats culture is even less flattering. For instance, in the first-season episode "Babylon,"[57] *Mad Men*'s protagonist Don Draper visits the Gaslight Café and encounters a number of countercultural characters who are, above all, even more self-absorbed and pretentious than himself. In fact, juxtaposed with overtly earnest Beatniks, the ironically detached advertising executive Draper emerges as the only truly cool character here. In the end, the scene segues from a folk band's performance of "Waters of Babylon" at the Gaslight into the episode's closing montage that transposes the song's lyrical message of being lost to other main characters of the show. Overall, this scene illustrates how *Mad Men* is yet another show that appeals to cool culture by prominently featuring alternative culture—without necessarily celebrating it uncritically—and simultaneously depicting it(self) as something different from the mainstream—yet not necessarily something outside of it.

Significantly enough, with Alexandra Patsavas, *Mad Men* featured the same music supervisor as *The O.C.*, as well as numerous other shows (e.g., *Grey's Anatomy*) and films (*Twilight*) that featured an indie soundtrack. Therefore, a closer look at these texts will provide us with further instances of the construction of alternative, "cool" culture as a collective realm through which it is possible—for television creators, audiences, and musical artists alike—to symbolically escape and oppose mainstream culture at the same time that this action is negotiated and problematized. However, this phenomenon is neither limited to the creative influence of Patsavas, nor is it restricted to explicitly youth-oriented genres, but has rather become a widespread element of contemporary Quality TV in general.

## Notes

1. Vodafone, "How Are You?" television advertisement, released 2001, video, 0:59, https://www.youtube.com/watch?v=YnZD2A47LbE.
2. The Dandy Warhols, "Bohemian Like You," track 10 on *Thirteen Tales from Urban Bohemia*, Capitol, 2000, compact disc.
3. Pyotr Ilyich Tchaikovsky, *Swan Lake*, composed in 1875-6.
4. Joseph Heath and Andrew Potter, *The Rebel Sell: How the Counterculture Became Consumer Culture* (Chichester: Capstone, 2005).
5. Dick Pountain and David Robins, *Cool Rules: Anatomy of an Attitude* (London: Reaktion Books, 2000), 19.
6. Michael Z. Newman, "Indie Culture: In Pursuit of the Authentic Autonomous Alternative," *Cinema Journal* 48, no. 3 (2009): 19.
7. Ben Aslinger, "*Nip/Tuck*: Popular Music," in *How to Watch Television*, ed. Ethan Thompson and Jason Mittell (New York: New York University Press, 2013), n.p.
8. Thomas Frank, *The Conquest of Cool: Business Culture, Counterculture, and the Rise of Hip Consumerism* (Chicago: The University of Chicago Press, 1997).
9. Ibid., 31.
10. Ibid., 26.
11. Ibid., 32.
12. Thomas Frank, "Alternative to What?" 1993, in *Commodify Your Dissent: Salvos from The Baffler*, ed. Thomas Frank and Matt Weiland (New York and London: W.W. Norton, 1997), 145.
13. Writing this chapter during the Trump presidency of course adds a whole different connotation to the idea of going against the grain of the mainstream. While there are indeed interesting resonances between these two discourses, I do not have the space here to analyze them at this point.
14. Heath and Potter, *The Rebel Sell*.
15. Richard Florida, *The Rise of the Creative Class* (New York: Basic Books, 2002); Jim McGuigan, *Cool Capitalism* (London: Pluto Press, 2009).
16. Andreas Reckwitz, *The Invention of Creativity*, 2012, trans. Steven Black (London: Polity Press, 2017), 5.
17. Often referred to as an "apparatus," Foucault's elusive concept is used by Reckwitz to describe something that "is not merely an institution, a closed functional system, a discourse or a set of values," but rather describes a socially and historically specific cultural formation that combines all of these elements and forces and "disposes people to a certain way of being." Reckwitz, *Invention of Creativity*, 28–9.
18. Ibid., 2.

19  *The O.C.*, season 2, episode 20, "The O.C. Confidential," dir. Tony Wharmby, feat. Adam Brody, Mischa Barton, and Ben McKenzie. Aired April 21, 2005, FOX. Warner Home Video, 2005, DVD.
20  *One Tree Hill*, season 1, episode 4, "Crash Into You," dir. David Carson, feat. Hilarie Burton, Chad Michael Murray, and James Lafferty. Aired October 14, 2003, The WB. Warner Home Video, 2005, DVD.
21  The Dandy Warhols, "We Used to Be Friends," track 2 on *Welcome to the Monkey House*, Capitol, 2003, compact disc.
22  The Decemberists, "The Infanta," track 6 on *Retrospective: The Music of Mad Men*, Republic Records, 2015, compact disc.
23  Red Hot Chili Peppers, *Californication*, Warner Bros., 1999, compact disc.
24  *The O.C.*, season 1, episode 15, "The Third Wheel," dir. Sandy Smolan, feat. Adam Brody, Mischa Barton, and Ben McKenzie. Aired January 7, 2004, FOX. Warner Home Media, 2004, DVD.
25  Beastie Boys, "Ch-Check It Out," track 1 on *To the 5 Boroughs*, Capitol, 2004, compact disc.
26  Various Artists, *Music from* The O.C.: *Mix 1-6*, Warner Bros., 2004–6, compact disc. For a more detailed account how transmedia phenomena like this feed into contemporary "convergence culture," see Henry Jenkins, *Convergence Culture: Where Old and New Media Collide* (New York: New York University Press, 2006).
27  Dorian Linskey, "How *The O.C.* Saved Music," *The Guardian*, December 1, 2004, https://www.theguardian.com/arts/features/story/0,11710,1363852,00.html.
28  Emily Zemler, "The O.C. Effect," *Popmatters*, January 13, 2005, https://www.popmatters.com/050114-indiesoundtracks-2496102955.html.
29  *The O.C.*, season 1, episode 2, "The Model Home," dir. Doug Liman, feat. Adam Brody, Mischa Barton, and Ben McKenzie. Aired August 12, 2003, FOX. Warner Home Video, 2004, DVD.
30  On the other hand, it is, of course, necessary to point out the irony of the Sex Pistols in this regard; while they have become one of the central icons of 1970s punk and origin point of the genre, they were nevertheless from the start much more of a cunning exploitation of the scene than an "authentically" subcultural phenomenon. Therefore, one could argue that the tensions I am outlining here were part of punk rock from its inception.
31  Aslinger, "*Nip/Tuck*," n.p.
32  Rachel Moseley, "The Teen Series," in *The Television Genre Book*, ed. Glen Creeber, 2nd ed. (London: BFI Publishing, 2008), 53–4.
33  *The O.C.*, season 1, episode 7, "The Escape," dir. Sanford Bookstaver, feat. Adam Brody, Mischa Barton, and Ben McKenzie. Aired September 16, 2003, FOX. Warner Home Media, 2004, DVD.
34  *The O.C.*, season 1, episode 21, "The Goodbye Girl," dir. Patrick Norris, feat. Adam Brody, Mischa Barton, and Ben McKenzie. Aired March 3, 2004, FOX. Warner Home Media, 2004, DVD.
35  Faye Woods, "Storytelling in Song: Television Music, Narrative and Allusion in *The O.C.*," in *Television Aesthetics and Style*, ed. Jason Jacobs and Steven Peacock (New York: Bloomsbury Academic, 2013), n.p.
36  Ibid., n.p.
37  Jeff Buckley, "Hallelujah," track 6 on *Grace*, Columbia, 1994, compact disc.

38  Woods, "Storytelling," n.p.
39  Frank, "Alternative to What?" 151–2.
40  See Frank Kelleter, "Five Ways of Looking at Popular Seriality," in *Media of Serial Narrative*, ed. Frank Kelleter (Columbus: The Ohio State University Press, 2017), 7–34; and Ruth Mayer, *Serial Fu Manchu: The Chinese Supervillain and the Spread of Yellow Peril Ideology* (Philadelphia, PA: Temple University Press, 2014).
41  Robert J. Thompson, *Television's Second Golden Age: From* Hill Street Blues *to* ER (Syracuse: Syracuse University Press, 1996); Amanda D. Lotz, *The Television Will Be Revolutionized*, 2nd ed. (New York and London: New York University Press, 2014).
42  As a matter of fact, the term was initially coined by Feuer, Kerr, and Vahimagi in 1984, but received widespread critical attention only after the publication of Thompson's study and its adaptation of the concept. See Jane Feuer, Paul Kerr, and Tise Vahimagi, eds., *MTM 'Quality Television'* (London: BFI Publishing, 1984).
43  Thompson, *Television's Second Golden Age*, 13.
44  Ibid., 14.
45  Michael Z. Newman and Elana Levine, *Legitimating Television: Media Convergence and Cultural Status* (New York and London: Routledge, 2012), 6–11.
46  Pountain and Robins, *Cool Rules*, 60–7.
47  *Californication*, season 3, episode 3, "Verities & Balderdash," dir. Davin Von Ancken, feat. David Duchovny, Natascha McElhone, and Pamela Adlon. Aired October 11, 2009, Showtime. Showtime, 2010, DVD.
48  *Californication*, season 2, episode 10, "In Utero," dir. Davin Von Ancken, feat. David Duchovny, Natascha McElhone, and Pamela Adlon. Aired November 30, 2008, Showtime. Showtime, 2009, DVD.
49  Nirvana, "Heart-Shaped Box," track 3 on *In Utero*, DGC, 1993, compact disc.
50  Newman, "Indie Culture," 16.
51  See also Kate Edenborg, "Going Groovy or Nostalgic: *Mad Men* and Advertising, Business, and Social Movements," in *Mad Men and Politics: Nostalgia and the Remaking of Modern America*, ed. Linda Beail and Lilly J. Goren (New York: Bloomsbury, 2015).
52  The Cardigans, "Great Divide," track 10 on *First Band on the Moon*, Stockholm Records, 1996, compact disc.
53  *Mad Men*, season 1, episode 2, "Ladies Room," dir. Alan Taylor, feat. Jon Hamm, Elizabeth Moss, and January Jones. Aired July 26, 2007, AMC. Lionsgate, 2008, DVD.
54  *Mad Men*, season 2, episode 6, "Maidenform," dir. Phil Abraham, feat. Jon Hamm, Elizabeth Moss, and January Jones. Aired August 31, 2008, AMC. Lionsgate, 2009, DVD.
55  The Decemberists, "The Infanta."
56  Charlie Wells, "'Mad Men' Is Set in the '60s, So Why Does It Use Music From Today?" *The Atlantic*, June 8, 2012, https://www.theatlantic.com/entertainment/archive/2012/06/mad-men-is-set-in-the-60s-so-why-does-it-use-music-from-today/258222/.
57  *Mad Men*, season 1, episode 6, "Babylon," dir. Andrew Bernstein, feat. Jon Hamm, Elizabeth Moss, and January Jones. Aired August 23, 2007, AMC. Lionsgate, 2008, DVD.

# Works Cited

Aslinger, Ben. "*Nip/Tuck*: Popular Music." In *How to Watch Television*, edited by Ethan Thompson and Jason Mittell, n.p. New York: New York University Press, 2013.
Beastie Boys. "Ch-Check It Out." Track 1 on To *the 5 Boroughs*. Capitol, 2004, compact disc.
Buckley, Jeff. "Hallelujah." Track 6 on *Grace*. Columbia, 1994, compact disc.
Edenborg, Kate. "Going Groovy or Nostalgic: *Mad Men* and Advertising, Business, and Social Movements." In *Mad Men and Politics: Nostalgia and the Remaking of Modern America*, edited by Linda Beail and Lilly J. Goren, 147–71. New York: Bloomsbury, 2015.
Feuer, Jane, Paul Kerr, and Tise Vahimagi, eds. *MTM "Quality Television"*. London: BFI Publishing, 1984.
Florida, Richard. *The Rise of the Creative Class*. New York: Basic Books, 2002.
Frank, Thomas. "Alternative to What?" 1993. In *Commodify Your Dissent: Salvos from* The Baffler, edited by Thomas Frank and Matt Weiland, 145–61. New York and London: W.W. Norton, 1997.
Frank, Thomas. *The Conquest of Cool: Business Culture, Counterculture, and the Rise of Hip Consumerism*. Chicago: The University of Chicago Press, 1997.
Hardwicke, Christine, dir. *Twilight*. 2008. Lionsgate, 2009. DVD.
Heath, Joseph, and Andrew Potter. *The Rebel Sell: How the Counterculture Became Consumer Culture*. Chichester: Capstone, 2005.
Jenkins, Henry. *Convergence Culture: Where Old and New Media Collide*. New York: New York University Press, 2006.
Kapinos, Tom, creator. *Californication*. 2007–2014. Showtime, 2015. DVD.
Kelleter, Frank. "Five Ways of Looking at Popular Seriality." In *Media of Serial Narrative*, edited by Frank Kelleter, 7–34. Columbus: The Ohio State University Press, 2017.
Linskey, Dorian. "How *The O.C.* Saved Music." *The Guardian*, December 1, 2004. https://www.theguardian.com/arts/features/story/0,11710,1363852,00.html.
Lotz, Amanda D. *The Television Will Be Revolutionized*. 2nd ed. New York and London: New York University Press, 2014.
Mayer, Ruth. *Serial Fu Manchu: The Chinese Supervillain and the Spread of Yellow Peril Ideology*. Philadelphia, PA: Temple University Press, 2014.
McGuigan, Jim. *Cool Capitalism*. London: Pluto Press, 2009.
Moseley, Rachel. "The Teen Series." In *The Television Genre Book*, edited by Glen Creeber, 52–4. 2nd ed. London: BFI Publishing, 2008.
Newman, Michael Z. "Indie Culture: In Pursuit of the Authentic Autonomous Alternative." *Cinema Journal* 48, no. 3 (2009): 16–34.
Newman, Michael Z., and Elana Levine. *Legitimating Television: Media Convergence and Cultural Status*. New York and London: Routledge, 2012.
Nirvana. "Heart-Shaped Box." Track 3 on *In Utero*. DGC, 1993, compact disc.
Pountain, Dick, and David Robins. *Cool Rules: Anatomy of an Attitude*. London: Reaktion Books, 2000.
Reckwitz, Andreas. *The Invention of Creativity*. Translated by Steven Black. 2012. London: Polity Press, 2017.
Red Hot Chili Peppers. *Californication*. Warner Bros., 1999, compact disc.
Rhimes, Shonda, creator. *Grey's Anatomy*. Buena Vista Home Entertainment, 2006–2020. DVD.

Schwahn, Mark, creator. *One Tree Hill*. 2003–2012. Warner Home Video, 2012. DVD.

Schwartz, Josh, creator. *The O.C.* 2003–2007. Warner Home Video, 2007. DVD.

Schwartz, Josh, and Stephanie Savage, creators. *Gossip Girl*. 2006–2012. Warner Home Video, 2013. DVD.

Star, Darren, creator. *Beverly Hills, 90210*. 1990–2000. Paramount, 2013. DVD.

The Cardigans. "Great Divide." Track 10 on *First Band on the Moon*. Stockholm Records, 1996, compact disc.

The Dandy Warhols. "Bohemian Like You." Track 10 on *Thirteen Tales from Urban Bohemia*. Capitol, 2000, compact disc.

The Dandy Warhols. "We Used to Be Friends." Track 2 on *Welcome to the Monkey House*. Capitol, 2003, compact disc.

The Decemberists. "The Infanta." Track 6 on *Retrospective: The Music of* Mad Men. Republic Records, 2015, compact disc.

Thomas, Rob, creator. *Veronica Mars*. Warner Home Video, 2005–2007. DVD.

Thompson, Robert J. *Television's Second Golden Age: From* Hill Street Blues *to* ER. Syracuse: Syracuse University Press, 1996.

Various Artists. *Music from* The O.C.: *Mix 1–6*. Warner Bros., 2004–6, compact disc.

Vodafone. "How Are You?" Television advertisement. Released 2001. Video, 0:59. https://www.youtube.com/watch?v=YnZD2A47LbE.

Weiner, Matthew, creator. *Mad Men*. 2007–2015. Lionsgate, 2008. DVD.

Wells, Charlie. "'Mad Men' Is Set in the '60s, So Why Does It Use Music From Today?" *The Atlantic*, June 8, 2012. https://www.theatlantic.com/entertainment/archive/2012/06/mad-men-is-set-in-the-60s-so-why-does-it-use-music-from-today/258222/.

Williamson, Kevin, creator. *Dawson's Creek*. 1998–2003. Sony Pictures Home Entertainment, 2006. DVD.

Woods, Faye. "Storytelling in Song: Television Music, Narrative and Allusion in *The O.C.*" In *Television Aesthetics and Style*, edited by Jason Jacobs and Steven Peacock, n.p. New York: Bloomsbury Academic, 2013.

Zemler, Emily. "The O.C. Effect." *Popmatters*, January 13, 2005. https://www.popmatters.com/050114-indiesoundtracks-2496102955.html.

# 9

# Sonic Sites of Subversion

## Listening and the Politics of Place in Karen Tei Yamashita's *Tropic of Orange*

Nathalie Aghoro

In Karen Tei Yamashita's *Tropic of Orange*, post-1992 Los Angeles comes to life as a globally connected and diverse city on the brink of a social and geopolitical disaster.[1] Yamashita's novel represents the metropolis as a complex site at once determined by overpowering visions of spectacle, Hollywood entertainment, and news images and by the ground-level abundance of sounds and noise that the city emits. The combined attention to the sights and sounds of the city constitutes a fictional space in the process of social transformation—a space in which sound registers the complex social, historical, and geographical conditions of change whose interconnections and repercussions are easily overlooked or remain invisible. By establishing a tangible sonic imaginary, *Tropic of Orange* captures the social dynamics of the city and connects the transnational reach of Los Angeles with the local fabric of its neighborhoods and the everyday experience of its residents.

Sound accompanies and shapes the perspectives of main characters who inhabit and tap into the city through listening. Manzanar Murakami, a former surgeon, lives in a shack but defies categorization, considering his life as a conductor on the freeway his vocation: "To say that Manzanar Murakami was homeless was as absurd as the work he chose to do. No one was more at home in L.A. than this man."[2] Social worker Buzzworm, dubbed "Angel of Mercy" or "walking social services," tirelessly pounds the streets of South Central to help others while listening to the radio on his Walkman and connecting to the people he encounters through what he hears.[3] The narrative enlists both for the configuration of alternative sonic maps of the city as they live and work in precarious social conditions.

Sound recording complements the emphasis on listening as acoustic activity in *Tropic of Orange* when television executive Emi and news reporter Gabriel Balboa—who is said to have an "almost audiographic memory" according to Buzzworm[4]—cross paths with the social worker and the homeless conductor. Their encounter engenders the takeover of the news cycle by marginalized voices and amplifies the acoustic ramifications of shifts in social dynamics.

I read Yamashita's literary engagement with the soundscapes of Los Angeles as a subversive sonic mapping of the city and its residents in print. As listeners and recorders, Emi, Buzzworm, Gabriel, and Manzanar connect to the sonic cues that the cityscape emits. They add their personal, auditory layers to the visual knowledge provided by traditional maps and take note of the sonic ripples in the fabric of the city's soundscape caused by the movement of people and current events. Not only do they act like cartographers who gather information about a place and its transnational, multiethnic formations to then produce a thematic map that reflects particular local features, they also subvert the entrenching effect of visual maps on power structures with the intimate knowledge of urban lifelines that their auditory awareness allows them to gather. At the same time, the novel points to the fallacies in sound recording, the transience of sound, and the futile effort of casting sound in writing. The combination of listening and recording characters as well as the references to media such as television, radio, newspapers, and their different ways of tapping into the sonic consistently resonate with the challenges of a novel that sets itself the task of channeling sound on the page.

## Politics of Place and Sonic Subversion

The novel's sonic cartographies explore the cultural and social stratifications of the urban environment and map them onto the auditory plane. The combination of sonic and spatial representation lays bare the political frictions in *Tropic of Orange*'s social imaginaries.[5] That these literary sounds of relational dynamics take place in the public sphere notably underscores the subversive role they play in Yamashita's fiction. The locations in which sounds are predominantly listened to and recorded are public places like the street, the freeway, the border, and the arena (a grand finale location dubbed the Pacific Rim Auditorium). As the plot unfolds, the spreading of sonic energy through and among these sites—expedited by media broadcasts on television, radio, and news reporting—conveys civic tensions with increasing intensity. Thus, *Tropic of Orange* demonstrates the significance of sound for social and political participation in a public sphere that it explicitly locates in place. For Jonathan Sterne, sound and the democratic public spere are closely interlinked. He argues that

> Habermas's notion of a public sphere—a complex concept that encompasses the process of democratic deliberation, the people who do it, and the space and conditions in which they come together—has a sonic dimension, from the centrality of speech to the problems of mass media, but its sonic component is rarely commented upon. All manner of publics have been constituted by audio media.[6]

The sonic cartographies of relational negotiations drive the social and political dynamics at play in *Tropic of Orange*. They situate the struggle for civic participation in a democratic context in places where the dominant public sphere encounters alternative publics of migrants and residents challenging the status quo. Sonic representation unfolds its constitutive power in the processes of resistance, renegotiation, and change shaping a social imaginary.

As people move from place to place or start using locations in unforeseen ways, processes of acoustic territorialization[7] are substantial for the claiming and reclaiming of places in Yamashita's fictionalized Los Angeles. The sonic dimension ascribed to the public sphere serves as a platform for the novel's concern with the politics of place in a local and global context.

Although Los Angeles is the place where the seven narrative strands of the novel converge, *Tropic of Orange* explicitly situates the city in the larger geographical area of the US-Mexican borderlands. One of the main storylines revolves around three characters on their way from Mexico to California dragging the tropic of cancer along with them, therewith disrupting the geographical shape of the city in a magical realist plotline.[8] Critics recognize in Yamashita's body of works a fundamental interest in what Jinqi Ling calls a "situated transnationality," which means that her novels represent the effects of global phenomena in specific places, emphasizing the particularities of a location while embedding it in an ever-evolving network of global relations.[9]

Indeed, the places in *Tropic of Orange*'s Los Angeles chosen by Manzanar and Buzzworm as primary sites for their vocational activities are characterized by a significant degree of mobility and fluidity. An eventful history of transitory territorial claiming and reclaiming defines Buzzworm's neighborhood and makes any attempt at mapping dangerously one-dimensional to say the least, if not downright futile. This becomes clear when he studies a map from "*Quartz City* or some such title"—an intertextual reference to Mike Davis's *City of Quartz* from 1990[10]: "He followed the thick lines on the map showing the territorial standing of Crips versus Bloods. Old map. 1972. He shook his head. Even if it were true. Even if it were true, whose territory was it anyway?"[11] What follows is a long list of possibilities to reconfigure the map, among them housing, marital status, ethnic composition, churches, ownership, and voter registration. In each instance, the resulting visualization remains incomplete and time-bound.[12] Buzzworm's reflections about the map pitch the liveliness and mutability of the streets he walks on against the utilitarian systematization of territorial claims. Read along the lines of Doreen Massey's conception of place, *Tropic of Orange* proposes a procedural conception of place. In "A Global Sense of Place" she writes, "First of all, it [place] is absolutely not static. If places can be conceptualized in terms of the social interactions which they tie together, then it is also the case that these interactions themselves are not motionless things, frozen in time. They are processes."[13] Buzzworm's question "whose territory was it anyway?"[14] contests the idea of defining place in terms of power and ownership. To do justice to the complex histories of suppression, destruction, and displacement associated with territorial claims, the novel resorts to soundmapping as a counterproject, an alternative form of sensory representation that captures places as processes in time and space. The Harbor freeway, Manzanar's site of musical work, exemplifies the scale that the novel ascribes to such a project. Like Buzzworm's "old hood,"[15] it is described in terms of permanence and mobility:

> Manzanar Murakami had become a fixture on the freeway overpass much like a mural or a traffic information sign or a tagger's mark. He was there every day, sometimes even when it rained, but it rarely rained. After all, this was L.A. There was a schedule of sorts, a program, an appropriate series of concerts and symphonies in accordance with the seasons and the climate of the city.[16]

Manzanar's presence on the overpass is a long-term process of becoming commonplace in a location exclusively geared toward movement. Like a fixture as persistent as a traffic information panel, he defies the infrastructure's innate discouragement of any prolonged stay. The comparison to murals and tags suggests that his art subverts the functional purpose of the traffic junction. The freeway is a "non-place"[17] par excellence, a site where people and goods pass through on their way to another destination. At the same time, Manzanar turns it into a location of artistic practice and musical creation that can only arise from a collective effort, showing very clearly that the global goes hand in hand with the local because of social interactions. "[S]tanding there, he bore and raised each note, joined them, united families, created a community, a great society, an entire civilization of sound. The great flow of humanity ran below and beyond his feet in every direction, pumping and pulsating, that blood connection, the great heartbeat of a great city."[18] Manzanar's compositions focus on the city's infrastructure as well as the motion patterns of its inhabitants. He conceives of the city as a large orchestra where different means of transportation, waterworks, power supplies, and telecommunication networks, among others, contribute their sound to a monumental musical performance. His perspective prioritizes social interactions among many individuals living in and moving through the same places as well as the global connections that these interactions entail. In contrast to the lack of representational depth that Buzzworm sees in maps, Manzanar's symphonies aspire to register and amplify local expressions of the global as they take place. His counterproject begins with the overpass and seeks to expand to connected destinations in the city and beyond on the wings of the sound waves in transit. The resulting procedural and multilayered musical cartography can be read as the fictional application of Massey's plea for "a sense of place, an understanding of 'its character,' which can only be constructed by linking that place to places beyond . . . a global sense of the local, a global sense of place."[19] In *Tropic of Orange*, sense refers not only to understanding but literally to the senses because Manzanar's overpass music dramatizes that knowing a city can be achieved by listening to its soundscape.

The novel's exploration of the ways in which sound and music take part in shaping social imaginaries in place does not end with sonic representations on the level of story. *Tropic of Orange* extends the attention to sound as literary form to narrative discourse as well. Yamashita complements the metaphorical dimension of the acoustic in narrative voice with a resonant language that codifies an aural, often even distinctly musical dimension into the act of narration. When asked about the musical inflections in the voices telling the novel's story, she imparts that she "worked a great deal at creating narrative voices that might capture this rhythmic sound sensibility"[20] and that "*Tropic* was a continuing experiment on narrative voice."[21] Seven distinct voices recount the seven narrative strands adding a unique storytelling vibe to each character's perspective. They are not only distinguishable in terms of sociolect, idiosyncratic markers, and third- or first-person narrative; they each also feature their own pacing, cadence, tonality, melody, and modulation. These acoustic features are tailored to each voice on the page through language and vocabulary selections, patterns of repetition, parallelisms, rhymes, and distinct sentence structures. In this way, the narrative voices acquire a sound register that lends an audible/readable fingerprint to each character's version of the story. Shifts from one combination of character perspective and narrative voice to another occur from chapter to chapter. The clear demarcation of sonic registers paves the way for overlaps in storytelling in-between chapters, revealing the overarching connections between characters and places. The multiple angles of *Tropic of Orange*'s narration mirror the relational focus in the configuration of its social imaginaries. Characters experience the same events from different points of view; conversations disclose the concerns and frictions that constitute the novel's public sphere; and shared locations become places of polyphonic encounter.

As a result, sound shapes *Tropic of Orange*'s politics of place in content and form. The narrative voices sound unique not least because the multiethnic cast of characters with different social backgrounds compels the protagonists to situate themselves in multiple places at once. In other words, Rafaela Cortes, Bobby Ngu, Emi, Buzzworm, Manzanar Murakami, Gabriel Balboa, and Arcangel[22] reconcile in themselves and among each other distinctive Los Angeles locations like Koreatown, Hollywood, South Central, and the Harbor freeway with Mazatlán, Singapore, México City, and the Pacific Rim. In her analysis of the novel, Melanie Pooch discusses the multidimensional approach to place as follows:

> The poetics of narrative, place, and code-switching equally contribute to the ethnic global city novel, displaying the multiperspectival and network-like structure, character relationships, and the interdependency of the "First world" and "Third world," converging in the simultaneous urban and border contact zone of the global(izing) city. As a result of the collaboration of the different characters, the seven individual worlds are revealed as one in the end, converging in "The World City." Yamashita thus creates a "third space" in and beyond the urban complexity of the cultural nodal point of Los Angeles, one in which interethnic identities are constantly negotiated beyond cultural lines.[23]

From a listening perspective, only an assembly of voices is capable of doing justice to the complex social dynamics constituting shared places. Song outlines that Yamashita's novel is a "literary attempt to capture a simultaneity of social experience"[24] and, one should add, that this endeavor is driven by the textualization of sound and the pronounced attention to sound and listening. *Tropic of Orange*'s encompassing sonic discourse is in sync with the interethnic and class-driven experiences of those who navigate and subvert social conditions locally on an everyday basis until their sounding grows in volume and breaches the privileged public sphere.

Sonic subversion arises in this literary framework specifically because—due to a familiar predominance of visual culture—sound is easily underestimated as a constitutive force of the social. The signal effects of sound are subject to a delay in social recognition and deploy their subversive force as long as they stay under the common perception threshold. This allows for the subversive potential of acoustic phenomena to become an agent for social change in the novel. As the following sections will show, the discrepancy between the emergence of soundscapes that resonate in and between places in *Tropic of Orange* and the time when they are registered in public perception creates a resounding field of social tension that holds the potential for future collective reconfigurations and simultaneously threatens to burst out in unrest and social upheavals.[25]

Overall, the novel's politics of place position individuals and their subjective, auditory experiences at the center of attention when it comes to understanding local relational interactions and their connection to global developments. The underlying assumption for the emphasis on the multiple character perspectives is that tangible knowledge of a place is acquired through situated sensory perception which, in turn, informs social relations. In *Tropic of Orange*, the processes of listening and soundmapping, in particular, put an emphasis on the dynamics of movement and communal change that resonate with the transience of sound.

## Transient Sounds and Orchestral Maps

The musical practice of conducting an orchestra is central to *Tropic of Orange*'s sonic engagement with the social. It combines two ways of listening that the conductor carries out simultaneously: Inward listening to compose, imagine, or project the desired musical composition and outward listening to take in external sonic impressions and unite the sounds of individual instruments into one musical performance. Conducting is thus a relational and intersubjective practice that encompasses the frictions in the novel's social imaginaries, rendering both the destructive and creative forces of the social in acoustic language and defining the social dynamics at play. The novel introduces Manzanar Murakami, accordingly, as a listening character who transcends and defies social boundaries.[26] As mentioned earlier, Manzanar is a homeless citizen of Los Angeles who stands on a freeway overpass and conducts an unlikely orchestra of street noise—capturing the rumble of trucks, the sound of sirens, and the roar of cars passing by with the movement of his conductor's baton transforming noise into classical music.[27] His musical orchestration of the freeway begins "within the very geology of the land" and extends to the "electric currents racing voltage into the open watts of millions of hungry energy-efficient appliances."[28] Manzanar's symphonic composition contains "the great overlay of transport—sidewalks, bicycle paths, roads, freeways, systems of transit both ground and air . . . variations both dynamic and stagnant, patterns and connections by every conceivable definition from the distribution of wealth to race, from patterns of climate to the curious blueprint of the skies."[29] The descriptive language in this passage fuses specialist terminologies from the fields of sociology, urban planning, musicology, and geography and therewith circumscribes the breadth and scale of Manzanar's soundmapping endeavor. He seeks no less than to incorporate the city's multiethnic and stratified formations as well as transnational currents and global connections in one monumental symphony. Jina Kim describes his mapping process as follows: "Manzanar's sonic composition transforms the static grid of the city into an interdependent ensemble of moving parts both material and human, a dynamic cartography that diverges from the superficial maps cluttering the novel."[30] The different movements and changing musical tempi in Manzanar's music reify the versatility and adaptability of his soundmapping as he works creatively with the variations in road capacity and usage on the freeway depending on the time of day and the current social climate: "For example, when the city rioted or when the city was on fire or when the city shook, the program was particularly apt, controversial, hair-raising, horrific, intense—apocalyptic, if you will. There was an incredibly vast repertory, heralding every sort of L.A. scenario."[31] Hence, the "Overture to the Santa Anas" differs substantially from "The Hour of the Trucks," a movement in which Manzanar dwells on "a noise that sounded like a mix of an elephant and the wail of a whale, concentrating until it moaned through the downtown canyons, shuddered past the on-ramps and echoed up and down the one-ten."[32] In Manzanar's case, soundmapping means the procedural attempt to project the noise of the city's material, social, and cultural multidimensionality into music.

Compared to spatiovisual maps, Manzanar's orchestration of place is an ambivalent topographical process with regard to questions of representational power. For Chiang, conducting—a practice that he defines as "a quintessentially nineteenth-century aesthetic practice"—represents an untimely approach that is bound to fail in a postmodern world: "In its modernist incarnation, the conductor is the figure capable of orchestrating and controlling the increasingly complex forms of modern social organization, but Murakami does not control anything. Rather, his task is sheer comprehension, and his conducting is an effort to grasp the conceptual order underlying the seemingly chaotic processes of the global system."[33] Nevertheless, the symphony takes shape and generates responses by an increasing number of people throughout the novel; therefore, it remains open whether he is in control of the orchestra or whether the assembly of infrastructural sounds controls his work. This ambivalence demonstrates that *Tropic of Orange* resists reestablishing a hierarchy of the senses between vision and sound. Instead, it proposes, as Esen Kara writes, that "Manzanar's imaginative mapping serves as an act of subversion against the homogenizing codes of conventional cartography."[34] Consequently, the novel represents the possibilities as well as the potential limitations of Manzanar's conducting and insistently frames soundmapping as an exploration in which the path forward is more consequential than the final destination and the search more important than certainty and control.

By choosing to adopt the perspective of a conductor instead of a musician who plays a particular instrument, *Tropic of Orange* conceives of listening as a dynamic and relational cultural practice. It is the active inquiry of a perceiving subject into a sonic environment. Attentive listeners in the novel are not just exposed to inner-city noise, but they are capable of (re)shaping the urban experience by paying attention to aesthetic values in the sonic output of a neoliberal system considered as nuisances or unwanted side effects. As Rodriguez observes, Manzanar's "position as a 'recycler,' and as a permanent inhabitant in social space, persisting in the public sphere as a man with no private home to return to or car to escape with, augments his perception of urban space, making visible a series of urban interconnections those 'on the map' can't see, since 'ordinary persons never bother to notice.'"[35]

Moreover, the novel suggests that listening to a particular place in addition to looking at it reveals the otherwise hidden repercussions of single events on urban dynamics. "As far as Manzanar was concerned," the narrative underlines, "it was all there. A great theory of maps, musical maps, spread in visible and audible layers—each selected sometimes purposefully, sometimes at whim, to create the great mind of music."[36] In keeping with the novel's focus on processes of marginalization and subversion, the narrative underscores Manzanar's underground status as an unheard artist: "Unknown to anyone, a man walking across the overpass at that very hour innocently hummed the recurrent melody of the adagio."[37] The narrative thus establishes a field of tension between his self-perception as a composer (and his previous profession as a successful surgeon) and the social categorizations generally applied to the homeless. Since Manzanar possesses neither an officially sanctioned place of residence nor regular employment, he is invisible to and silenced by society. Only very few people pay attention to his activities on the overpass, and, if they do, they rather tend to question his sanity. As a result, his sonic cartographies convey an expansive fictional map of the city to the reader, but they remain largely underappreciated in the storyworld because of the sounds he works with.

## Changing the Tune: Sonic Subversion and Social Change

The simultaneity of voices—each narrating a chapter per day—dramatizes the discrepancy between individual and alternative approaches to place. The change from one sonic register to another fans out the plurality of situated soundscapes and demonstrates a participatory mode for representing place. "*There are maps and there are maps and there are maps*" in *Tropic of Orange*,[38] pointing toward the existence of many possible ways to map the multiple layers that constitute Los Angeles, even (and, one could argue, especially) in times of social unrest and transformation. In her cultural history about Los Angeles, Gaye Theresa Johnson argues that "[s]truggles for social justice in Los Angeles involved changing the meaning of existing spaces and creating new ones."[39] She defines the "way in which marginalized communities have created new collectivities based not just upon eviction and exclusion from physical places, but also on new and imaginative uses of technology, creativity, and spaces" as "spatial entitlement."[40] The development of "shared soundscapes" acquires heightened significance for young people of color, according to Johnson, when "housing segregation, police containment, and transit racism ... [make] it difficult to move across urban spaces."[41] *Tropic of Orange* fictionalizes these sonic cultures of resistance in its literary discourse on claiming and reclaiming places. As Kara describes it, "[t]he claim for the right to the city, first of all, manifests itself in the novel as the right to create your own map of the city."[42] The metadiscursive map in the frontmatter of *Tropic of Orange*, entitled HyperContexts,[43] visualizes the participatory and equal distribution of voices and perspectives for each day of the week: "This spatial structure of the narrative echoes the multiple rhythms of the city that have been made silent by the constellation of power and geographical knowledge. The novel thus questions the one-dimensionality of map-making as an ideological strategy at the hands of power."[44] The novel's counternarrative to maps as tools for imperialist land grab and the hegemonial assertion of power reenvisions mapping as a sonic practice by deploying different possibilities of sound recording and listening. While Manzanar's soundmapping consists of conducting from a stationary, overview position, Buzzworm listens to the city from the ground level as he remains constantly on the move. As a consequence, their coexisting soundmaps differ because their respective ways of being in the world fundamentally shape their approaches to place and sound.

Buzzworm projects a pedestrian map of Los Angeles—which in itself can be read as the resistant and potentially dangerous act of a Black protagonist. As he walks on the streets offering his help to others, he constantly listens to the radio on his Walkman. However, his listening attention is partitioned: he has "sound plugged into one ear" and he keeps the other ear open for the people who approach him on his rounds.[45] Thus, he gathers information on current events on the radio and, at the same time, knows what preoccupies the residents in the neighborhood. This dual receptivity gives him the ability to match the two kinds of knowledge that he acquires in order to deduce, for example, where social tensions could possibly arise or how City Hall decisions on urban planning affect the inhabitants, most of whom he knows by name. Because of his unique knowledge, representatives of the news media repeatedly seek him out if they need information on what is going on in the city. Activists and politicians, "big guns" as he calls them, also contact him "every time there's trouble" and ask for his help when they want to meet the community to "preach ... the gospel of hope not dope."[46]

Buzzworm's walking perspective zooms in on the diverse details that constitute the urban experience. He maps the individual fates of inhabitants in the area through the soundscape that they produce. The Walkman as his constant companion helps him to merge the sounds of the sidewalk with the invisible space of radio broadcasting deployed as a socially stratified, airborne sound layer over the city and, thus, to stay in tune with the sounds and rhythm of the city. For Buzzworm, his mobile connection to the radio serves to keep him in touch with the city's multilayered soundscapes:

> Twenty four hours, Buzzworm was listening to the radio. From station ID to station ID. Unless he meant business, he had it plugged in like supermarket music, just in the background to help you shop, give a little light rhythm to the situation ... And he listened to everything. He listened to rap, jazz, R&B, talk shows, classical, NPR,

religious channels, Mexican, even the Korean channel. Didn't know a thing they were saying, but he liked the sounds. Fact is, he listened to the sounds so much he could imitate them.[47]

Buzzworm's listening practice resists and actively counters the systemic segregation of communities by persistently staying in touch with the multiplicity of musical styles, sounds, and voices in the air. Along the lines of Michelle Hilmes, who argues that radio is a medium that constructs imagined communities,[48] he maps the social relations in the city along the parameters of place and sound as he listens to the radio while walking through the city. For Jean-Paul Thibaud, "[u]sing a Walkman in public places is part of an urban tactic that consists of decomposing the territorial structure of the city and recomposing it through spatio-phonic behaviours. Double movement of deterritorialization and reterritorialization."[49] Buzzworm's connected listening allows him to build two simultaneous connections, to the people on the streets and to the music and public broadcasts on the channel he actively selects. It illustrates the discrepancy between the social constrictions and possible communal futures that he imagines with every step he takes. His sonic self-determination subverts the institutional power of city officials, investors, and urban planners and articulates the potential for an alternative politics of place.

By establishing Buzzworm as a connected listener, *Tropic of Orange* suggests that paying attention to the sounds of the city and listening to the various musical styles moving the residents and defining their sonic relation to place is an efficient way of uniting the community not into a unitary, but diversified whole with various cultures enriching the unique, lived experience of a place. As Rodriguez argues, "The value Buzzworm places upon his local environs, whether by admiring the unique attributes of the palms lining his streets, or by providing 'walking social services' in the absence of government aid, suggests the possibility of a counter-map, one created from the ground up."[50] Buzzworm deliberately advocates his soundmapping approach as he recommends music to the people he meets in his everyday encounters, such as the immigrant street peddlers, most of them without papers, who "had their unspoken territories, too," just like everyone else in the city.[51] When one of them, Margarita, asks him "What's the music today?"[52] he responds:

"For you, Margarita, oldies."
"Aretha Franklin."
"How'd you know?"
"I know. I know you listen to mariachi tambien."
"Los Camperos. The very best."
"Sorry," she shook her head. "It is not my culture. I, Salvador."
"You," he pointed at her. "Aretha Franklin. Don't be such a purist."[53]

The conversation shows that he seeks to actively foster solidarity between marginalized social groups in his city through sound and music. Buzzworm advocates listening to each other in order to counter the divisive tactics of domination with which "some wanted to pit black against brown."⁵⁴ Consequently, social change means for him to "get behind another man's perspectives. Hear life in another sound zone. Walk to some other rhythms."⁵⁵ The combination of movement and sound mirrors the intersubjective openness and flexibility that characterizes Buzzworm's perspective.

Although their approaches to shaping the city through sound and listening differ fundamentally, Manzanar and Buzzworm both project the city as a transitory and fluid place with their soundmaps. Sound reflects the urban characteristic of permanent change because it is able to create, as LaBelle writes, "a relational space, a meeting point . . . a geography of intimacy that also incorporates the dynamics of interference, noise, transgression."⁵⁶ Since both characters are attuned to the urban soundscape, they are also the ones who register the interferences caused by the three entangled disasters that hit the city: Oranges spiked with cocaine, the northward displacement of the tropic of cancer,⁵⁷ and the violent military crackdown on an alternative community taking up residence on the freeway after a monumental car accident.

## The Potentials and Limitations of Sonic Subversion

Sonic subversion requires a broad audience to reach momentum in Yamashita's Los Angeles. Media recordings and broadcasts take on the role of amplifiers for Buzzworm's and Manzanar's sound projections of equitable and encompassing social imaginaries. When their storylines intersect with news reporter Gabriel and television executive Emi, their sonic subversions cross the threshold to mainstream media and breach the public sphere. A broader public discourse on social relations ensues (and ultimately attracts the attention of institutional forces seeking to maintain the status quo).

Slowly and almost unnoticed, Manzanar's symphony impacts people and places, first within earshot, then through the sound fabric that constitutes the urban soundscapes in *Tropic of Orange*. Unconsciously, a rising number of people tune in on his wavelength. When Buzzworm arranges an interview with Manzanar for Gabriel, the news reporter's acoustic encounter with the conductor precedes his first sighting of the older man on the overpass:

> I walked west with some urgency, determined to find my subject . . . A sooty heat and din emanated from there, pressed against what I imagined to be all the elastic parts of my body: my lungs, my diaphragm, my tympanum.
> 
> And as I looked across, I saw him. Buzzworm was right. There he was larger than life . . . his arms reaching and caressing the air for the sound and rhythms of . . . of what?⁵⁸

Gabriel hears, but does not understand or recognize Manzanar's music, although it reaches him in the organs in which resonance, speech, and hearing thrive. He is attracted to his music, but cannot listen to the symphony consciously since the sounds and noises Manzanar works with are not part of his musical listening repertoire. Even after Murakami explains his creative process, Gabriel sees a homeless man first and foremost and, as a consequence, doubts his words. When he tells Emi about the encounter later on, he cannot draw the connection just yet and is therefore unable to connect his observation about a choir on the freeway to Manzanar Murakami: "They're all singing, humming. I mean it's sporadic, but yeah. Homeless singing, harmonizing. Something."[59] However, Gabriel is aware of his incapacity to register what he hears because of his preconceptions: "Buzzworm was right. There was something important about this man, so wise, so completely honest. He deserved my respect . . . And there I was without a pen or an audio recorder, without words."[60] His writing can therefore be merely a reconstruction of the interview that he failed to record on site.[61] His plight points to the transience of sound as it challenges the notion that writing could replace orality entirely as archive and reinforces the emphasis on the importance of acknowledging a plurality of perspectives when recording social configurations in place.

Overall, the lack of recognition of Manzanar's work despite its tangible impact on others establishes a metadiscourse on the social invisibility and the marginalization of inhabitants without an official address that is sustained by the novel's exploration of connected listening. To further his agenda of social change based on solidarity and shared soundscapes that transcend class and ethnicity, Buzzworm convinces local television executives to set up a show that reports from the settlement consisting of abandoned cars on the freeway in the aftermath of the rush hour accident. Emi becomes the producer of "What's The Buzz?" a show in which the homeless who experiment with alternative ways of living in the city—including, for instance, urban gardening in one of the immobilized cars—share their thoughts, experiences, and perspectives on the city's social issues.[62] Eventually, the show's intervention in public discourse is cut short by commercials and the military, but despite the violent dissolution of the situation, the message of possible counternarratives to the dominant order endures.

Sonic subversion in *Tropic of Orange* challenges hegemonial and neoliberal politics of place. In the novel, sound represents a transgression and a subversion of social and political forces that seek to delineate fixed boundaries and shut themselves off from those that they perceive as fundamentally different. *Tropic of Orange* points to both the potentials and limitations of sonic subversion and listening practices as it textualizes the simultaneity and multiplicity of the social through the situated perspectives of listeners and recorders of sound.

## Notes

1   Although Yamashita's 1997 novel never explicitly mentions the 1992 Los Angeles protests following the acquittal of police officers involved in the beating of Rodney King, police violence and the ensuing events can be read as a marked absence, or, in acoustic terms, a silent resonance in *Tropic of Orange*. Lynn Mie Itagaki understands *Tropic of Orange* accordingly as "a parable of the everyday acts of the oppressed, illegal, invisible, forgotten, and powerless that nonetheless influence and shape the world. Contesting the racist practices of racial civility, Yamashita depicts the imaginative and often subtle ways in which the less powerful resist the demands of

the more powerful" (Lynn Mie Itagaki, *Civil Racism: The 1992 Los Angeles Rebellion and the Crisis of Racial Burnout* [Minneapolis: University of Minnesota Press, 2016], 31). The disaster that strikes in *Tropic of Orange* can thus be understood as magical realist processing of the 1992 events that had a lasting effect on the political, social, and cultural discourses about the city.

2 Karen Tei Yamashita, *Tropic of Orange* (Minneapolis, MN: Coffee House Press, 1997), 36.
3 Ibid., 26.
4 Ibid., 108.
5 On social imaginaries, see: Suzi Adams, Paul Blokker, Natalie J. Doyle, John W. M. Krummel, and Jeremy C. A. Smith, "Social Imaginaries in Debate," *Social Imaginaries* 1, no. 1 (2015): 15–52; and Charles Taylor, *Modern Social Imaginaries* (Durham, NC: Duke University Press, 2003).
6 Jonathan Sterne, "Collectivities and Couplings," in *The Sound Studies Reader*, ed. Jonathan Sterne (London: Routledge, 2012), 325–6.
7 Brandon LaBelle's notion of "acoustic territories" that he introduces to the field of auditory culture (and that I discuss in the introduction to this edited collection) is helpful for the conceptualization of the soundmapping processes that take place in Yamashita's fiction. For LaBelle, the soundscapes one encounters in a place are imbued with cultural and political agency and this dynamic characteristic substantially drives the narrative discourse as well as the action of the plot in the novel. LaBelle argues that "[t]he seemingly innocent trajectory of sound as it moves from its source and toward a listener, without forgetting all the surfaces, bodies, and other sounds it brushes against, is a story imparting a great deal of information fully charged with geographic, social, psychological, and emotional energy" (Brandon LaBelle, *Acoustic Territories: Sound Culture and Everyday Life* [New York: Continuum, 2010], xvi). The aim of this chapter is to show that in *Tropic of Orange* sound not only provides world building information, but also motivates the characters' social behavior and weaves a conceptual thread with political force through the distinct storylines on the level of narration.
8 On the topic of magical realism and migration at the US-Mexican border in *Tropic of Orange*, see: Hande Tekdemir, "Magical Realism in the Peripheries of the Metropolis: A Comparative Approach to *Tropic of Orange* and *Berji Kristin: Tales from the Garbage Hills*," *The Comparatist* 35 (2011): 40–54 and "Post-Frontier and Re-definition of Space in *Tropic of Orange*," in *Blast, Corrupt, Dismantle, Erase: Contemporary North American Dystopian Literature*, ed. Brett Joseph Grubisic, Gisèle M. Baxter, and Tara Lee (Wilfrid Laurier University Press, 2014), 93–110.
9 Jinqi Ling, *Across Meridians: History and Figuration in Karen Tei Yamashita's Transnational Novels* (Stanford, CA: Stanford University Press, 2012), 25. Min Hyoung Song observes that the novel "imagines Los Angeles as a meeting place between the national and global" (Min Hyoung Song, "Becoming Planetary," *American Literary History* 23, no. 3 [2011]: 556). In *A Body of Individuals*, Sue-Im Lee delineates the "phenomena of globalization" that Yamashita's fiction addresses: "[T]he high-speed information, media, and transportation technologies; the transnational modes of production and consumption; the accelerated flow of people, capital, goods, information, and entertainment; all of which result in the shift in the human experience of space, distance, and time" (Sue-Im Lee, *A Body of Individuals:*

*The Paradox of Community in Contemporary Fiction* [Columbus: Ohio State University Press, 2009], 62).
10 Mike Davis, *City of Quartz: Excavating the Future in Los Angeles* (New York: Verso, 2006).
11 Yamashita, *Tropic of Orange*, 80–1.
12 Buzzworm lists the various ways of mapping the area when news reporter Balboa asks him to determine the accuracy of gang territory. According to Kevin Cooney, "[o]bjecting to what the map leaves out, Buzzworm critiques Gabriel's assumption that gang activity is the only sociological criterion for mapping Compton . . . In listing alternative ways of mapping the neighborhood, Buzzworm's questions move beyond the media's sensational focus on gang rivalries and call attention to the lived experience of ordinary residents" (Kevin Cooney, "Metafictional Geographies: Los Angeles in Karen Tei Yamashita's *Tropic of Orange* and Salvador Plascencia's *People of Paper*," in *On and Off the Page: Mapping Place in Text and Culture*, ed. M. B. Hackler [Newcastle upon Tyne: Cambridge Scholars Publishing, 2009], 197).
13 Doreen Massey, "A Global Sense of Place," in *Space, Place, and Gender* (Minneapolis: University of Minnesota Press, 1994), 155.
14 Yamashita, *Tropic of Orange*, 81.
15 Ibid., 80.
16 Ibid., 36.
17 On Marc Augé's notion of "non-place," see: Marc Augé, *Non-Lieux: Introduction à une anthropologie de la surmodernité* (Paris: Éditions du Seuil, 1992).
18 Yamashita, *Tropic of Orange*, 35.
19 Massey, "A Global Sense of Place," 156.
20 Elizabeth P. Glixman, "An Interview with Karen Tei Yamashita," *Eclectica* 11, no. 4 (2007): n.p. http://www.eclectica.org/v11n4/glixman_yamashita.html.
21 A. Robert Lee, "Speaking Craft: An Interview with Karen Tei Yamashita," in *Karen Tei Yamashita: Fictions of Magic and Memory*, ed. A. Robert Lee (Honolulu: University of Hawai'i Press, 2019), 183.
22 The names of the seven protagonists in *Tropic of Orange* as they appear in the "HyperContexts" positioned after the table of contents and the novel itself.
23 Melanie Pooch, *DiverCity – Global Cities as a Literary Phenomenon: Toronto, New York, and Los Angeles in a Globalizing Age* (Bielefeld: Transcript Verlag, 2016), 177.
24 Song, "Becoming Planetary," 558.
25 Rodriguez defines the tensions in *Tropic of Orange* in spatial terms, using Henri Lefebvre's concept of a subversive "differential space" between lived, perceived, and conceived space. She argues that "[t]his concept of differential space, as a subversive force that accentuates differences and transgresses against a given spatial practice, is directly applicable to *Tropic of Orange*. At a metanarrative level, the term describes the imaginative geography Yamashita creates in the novel itself, as her non-linear, non-hierarchical, polyvocal storylines produce a literary differential space. At the narrative level . . . several of the central events of *Tropic of Orange*, including the homeless takeover of the Harbor freeway and the citywide traffic symphony, are best understood as expressions of differential space" (Cristina M. Rodriguez, "'Relentless Geography': Los Angeles' Imagined Cartographies in Karen Tei Yamashita's *Tropic of Orange*," *Asian American Literature: Discourses and Pedagogies* 8 [2017]: 127–8).

From the perspective of listening and auditory culture, the present essay carves out the combined significance of sound and place for the literary rendering of the differential space that Rodriguez recognizes in Yamashita's work.

26  For more on the historical meaning of place and name in Manzanar Murakami's connection to the Manzanar internment camp, see: Gayle K. Sato, "Post-Redress Memory: A Personal Reflection on Manzanar Murakami," *Concentric: Literary and Cultural Studies* 39, no. 2 (2013): 119–35.
27  Mark Chiang speaks of "musique concrete" in reference to Manzanar's oeuvre (Mark Chiang, "Capitalizing Form: The Globalization of the Literary Field: A Response to David Palumbo-Liu," *American Literary History* 20, no. 4 [2008]: 841).
28  Yamashita, *Tropic of Orange*, 57.
29  Ibid.
30  Jina B. Kim, "Toward an Infrastructural Sublime: Narrating Interdependency in Karen Tei Yamashita's Los Angeles," *MELUS* 45, no. 2 (2020): 1.
31  Yamashita, *Tropic of Orange*, 36.
32  Ibid., 36, 119.
33  Chiang, "Capitalizing Form," 841–2.
34  Esen Kara, "Rewriting the City as an Oeuvre in Karen Tei Yamashita's Tropic of Orange," *Interactions: Ege Journal of British and American Studies* 27 (2018): 79.
35  Rodriguez, "'Relentless Geography,'" 126. Here, she directly refers to the passage on page 57 in *Tropic of Orange*.
36  Yamashita, *Tropic of Orange*, 57.
37  Ibid.
38  Ibid., 56.
39  Gaye Theresa Johnson, *Spaces of Conflict, Sounds of Solidarity: Music, Race, and Spatial Entitlement in Los Angeles* (Berkeley: University of California Press, 2013), xii.
40  Ibid., x.
41  Ibid., xiii.
42  Kara, "Rewriting the City as an Oeuvre," 78.
43  Yamashita, *Tropic of Orange*, n.p.
44  Kara, "Rewriting the City as an Oeuvre," 78.
45  Yamashita, *Tropic of Orange*, 27.
46  Ibid., 216.
47  Ibid., 29.
48  Michelle Hilmes, "Radio and the Imagined Community," in *The Sound Studies Reader*, ed. Jonathan Sterne (London: Routledge, 2012), 351.
49  Jean-Paul Thibaud, "The Sonic Composition of the City," in *The Auditory Culture Reader*, ed. Michael Bull and Les Back (Oxford: Berg, 2006), 329.
50  Rodriguez, "'Relentless Geography,'" 123.
51  Yamashita, *Tropic of Orange*, 84.
52  Ibid.
53  Ibid.
54  Ibid., 102.
55  Ibid., 103.
56  LaBelle, *Acoustic Territories*, xvi–xvii.
57  The earth literally shifts under their feet when the tropic arrives in Los Angeles, and while Manzanar feels the vibrations with his body and incorporates them into his

symphony, Buzzworm hears that "radio stations on every dial [are] holding their notes, their words, their voices, their dead air" (Yamashita, *Tropic of Orange*, 137).
58  Yamashita, *Tropic of Orange*, 46.
59  Ibid., 155.
60  Ibid., 108.
61  Ibid.
62  Ibid., 175–80.

## Works Cited

Adams, Suzi, Paul Blokker, Natalie J. Doyle, John W. M. Krummel, and Jeremy C. A. Smith. "Social Imaginaries in Debate." *Social Imaginaries* 1, no. 1 (2015): 15–52.

Augé, Marc. *Non-Lieux: Introduction à une anthropologie de la surmodernité*. Paris: Éditions du Seuil, 1992.

Chiang, Mark. "Capitalizing Form: The Globalization of the Literary Field: A Response to David Palumbo-Liu." *American Literary History* 20, no. 4 (2008): 836–44.

Cooney, Kevin. "Metafictional Geographies: Los Angeles in Karen Tei Yamashita's *Tropic of Orange* and Salvador Plascencia's *People of Paper*." In *On and Off the Page: Mapping Place in Text and Culture*, edited by M. B. Hackler, 189–218. Newcastle upon Tyne: Cambridge Scholars Publishing, 2009.

Davis, Mike. *City of Quartz: Excavating the Future in Los Angeles*. New York: Verso, 2006.

Glixman, Elizabeth P. "An Interview with Karen Tei Yamashita." *Eclectica* 11, no. 4 (2007): n.p. http://www.eclectica.org/v11n4/glixman_yamashita.html.

Hilmes, Michelle. "Radio and the Imagined Community." In *The Sound Studies Reader*, edited by Jonathan Sterne, 351–62. London: Routledge, 2012.

Itagaki, Lynn Mie. *Civil Racism: The 1992 Los Angeles Rebellion and the Crisis of Racial Burnout*. Minneapolis: University of Minnesota Press, 2016.

Johnson, Gaye Theresa. *Spaces of Conflict, Sounds of Solidarity: Music, Race, and Spatial Entitlement in Los Angeles*. Berkeley: University of California Press, 2013.

Kara, Esen. "Rewriting the City as an Oeuvre in Karen Tei Yamashita's Tropic of Orange." *Interactions: Ege Journal of British and American Studies* 27 (2018): 75–89.

Kim, Jina B. "Toward an Infrastructural Sublime: Narrating Interdependency in Karen Tei Yamashita's Los Angeles." *MELUS* 45, no. 2 (2020): 1–24.

LaBelle, Brandon. *Acoustic Territories: Sound Culture and Everyday Life*. New York: Continuum, 2010.

Lee, A. Robert. "Speaking Craft: An Interview with Karen Tei Yamashita." In *Karen Tei Yamashita: Fictions of Magic and Memory*, edited by A. Robert Lee, 177–88. Honolulu: University of Hawai'i Press, 2019.

Lee, Sue-Im. *A Body of Individuals: The Paradox of Community in Contemporary Fiction*. Columbus: Ohio State University Press, 2009.

Ling, Jinqi. *Across Meridians: History and Figuration in Karen Tei Yamashita's Transnational Novels*. Stanford, CA: Stanford University Press, 2012.

Massey, Doreen. "A Global Sense of Place." In *Space, Place, and Gender*, 146–56. Minneapolis: University of Minnesota Press, 1994.

Pooch, Melanie. *DiverCity – Global Cities as a Literary Phenomenon: Toronto, New York, and Los Angeles in a Globalizing Age*. Bielefeld: Transcript Verlag, 2016.

Rodriguez, Cristina M. "'Relentless Geography': Los Angeles' Imagined Cartographies in Karen Tei Yamashita's *Tropic of Orange.*" *Asian American Literature: Discourses and Pedagogies* 8 (2017): 104–30.

Sato, Gayle K. "Post-Redress Memory: A Personal Reflection on Manzanar Murakami." *Concentric: Literary and Cultural Studies* 39, no. 2 (2013): 119–35.

Song, Min Hyoung. "Becoming Planetary." *American Literary History* 23, no. 3 (2011): 555–73.

Sterne, Jonathan. "Collectivities and Couplings." In *The Sound Studies Reader*, edited by Jonathan Sterne, 325–8. London: Routledge, 2012.

Taylor, Charles. *Modern Social Imaginaries*. Durham, NC: Duke University Press, 2003.

Tekdemir, Hande. "Magical Realism in the Peripheries of the Metropolis: A Comparative Approach to *Tropic of Orange* and *Berji Kristin: Tales from the Garbage Hills.*" *The Comparatist* 35 (2011): 40–54.

Tekdemir, Hande. "Post-Frontier and Re-definition of Space in *Tropic of Orange.*" In *Blast, Corrupt, Dismantle, Erase: Contemporary North American Dystopian Literature*, edited by Brett Joseph Grubisic, Gisèle M. Baxter, and Tara Lee, 93–110. Waterloo: Wilfrid Laurier University Press, 2014.

Thibaud, Jean-Paul. "The Sonic Composition of the City." In *The Auditory Culture Reader*, edited by Michael Bull and Les Back, 329–14. Oxford: Berg, 2006.

Yamashita, Karen Tei. *Tropic of Orange*. Minneapolis, MN: Coffee House Press, 1997.

# From "Dead Spots" to "Hot Spots"
## Ann Petry's "On Saturday the Siren Sounds at Noon"[1]

Jennifer Lynn Stoever

In March and April 2020, think piece after think piece went live witnessing the dramatic increase in the sound of sirens in New York City's streets due to the swift, deadly, and downright horrific outbreak of Covid-19. From tabloids (*New York Post*) to venerable source (*New York Times*), local sites (*Gotham Gazette*) to national outlets (*The New Yorker*), fashion-forward glossy (*Vogue*) to political monthly (*New Republic*), New York City–centric stories resounded with the "relentless," "often-interminable," "unnervingly frequent," "dirge" of sirens—the "ambulance's horrid solo"—coupled with an "eerie quietness in between them."[2] Articles like these, often with poetic and ominous titles such as "To, and Beyond, the Siren's Call" and "Alone in the City of Sirens" show members of the city's professional creative class struggling to document, memorialize, and make sense of the unknown terror of the novel virus through a mutually constitutive engagement with the city soundscape, all the louder given Governor Andrew Cuomo's mandated state closure and self-isolation period that left New York otherwise muted.[3] As Molly Jong-Fast described for *Vogue*, the sirens marked New York City as it became "the epicenter of the epicenter . . . the ground zero of death."[4] As I struggle to write this chapter, the coronavirus pandemic has taken the lives of 2,042,200 people globally, 400,000 people in the United States and over 26,000 residents of New York City alone.[5]

Throughout the crisis, the state has insisted that a preexisting American "we" faced Covid-19 together, and that the same rules and vulnerabilities apply to all. Governor Cuomo warned in one of his daily press briefings, on March 31, 2020, "this virus does not discriminate—no one is immune to it—and people must continue to be cautious."[6] While technically true, statements like these gloss over the human-made disasters of racism, economic oppression, and medical apartheid that already made some populations far more vulnerable than others, even before the virus's arrival. "If you are Black, Brown, or poor, you are disproportionately over-represented in Covid-related death, hospitalization, or infections," reported the Covid Policing Project.[7] Some news outlets appealed to the sound of sirens as a unifying force in and of itself. The Associated Press, for example, contrasted the increase of sirens in New York City with a growing quiet enabling residents to notice more birdsong, captioning its April

11, 2020, coronavirus video: "[I]n the grind of the pandemic, sound has become one of our shared experiences."[8] Yes . . . and no. Though most people have the potential to hear sirens in some fashion, whether as a "mournful electric coyote" in the ear and/or deeper vibrations in the feet, bones, and body, residents do not all hear and react to them in remotely the same way.[9] Birdsong, for that matter, didn't automatically create shared experiences for New Yorkers either, as birder and science enthusiast Christian Cooper's cell footage attested a month into lockdown, recorded in self-protection in Central Park's Ramble area after a white woman named Amy Cooper called the police, falsely claiming "an African American man [is] threatening my life" after he had asked her to leash her dog as the rules stated, so as not to disturb the birds.[10] Amy Cooper's call was intended to send police sirens to the scene, and could have led to Christian Cooper's death.

As the tale of two Coopers illustrated, yet again, that while the sirens may sound *to* all, they most definitely do not sound *for* all, and there were very divergent descriptions of and reactions to New York's Covid sirens as well. *Times* writer Zoladz realized she "was a bit ashamed" to have been living only a mile from the Brooklyn Hospital Center for over five years, yet had comfortably relegated the sirens to background city noise until March 2020.[11] "I am fortunate to have until now," Zoladz said, "to have moved through life with a breezy ignorance of the nearest hospital's location." Now she notes that "to listen to an ambulance siren is to picture the face and the body and the family of the person it is carrying to the hospital."[12] In *The New Republic*, Samer Kalaf, too, articulated his past and present privilege by fixating "on a sound associated with emergency that is no longer reduced to just another facet of the city's ambient blare" to take note that "a person is dying *out there*."[13] For Kalaf, to be alive, well, and listening in his apartment means using his platform to advocate for those who are not. His "process does not end when the siren does. An idle mind might consider the overcrowded emergency rooms, the underpaid and underprotected hospital personnel, the American health care system that was already broken."[14] New Yorkers' mental, emotional, and economic states shaped how they listened to the sounds of early Covid lockdown, and hearing these sounds—at once very familiar *and* radically changed—greatly impacted their understanding of themselves and their positions in the social world they were now inhabiting.

In a systemically overwhelming crisis such as the global Covid pandemic, sirens sound out listeners' social positions as much as they alert them to the emergencies at hand, particularly in a nation like the United States, structured by long, violent histories of hierarchical race, class, and gender divisions that determine whose lives matter and whether or not one gets a chance to live at all. I intentionally echo the rallying cry of the Black Lives Matter movement here because the police sirens that attended summer 2020—those that initially attended the murder of George Floyd by Minneapolis police officer Derek Chauvin in May and those that met the bold waves of masked and socially distanced protesters for racial justice in New York City and across the country—sound an interconnected part of America's ongoing crisis. The same communities still facing ongoing police violence and surveillance are dying of Covid-19 in outsized numbers relative to white Americans: at the rate of three to one, according to the National Urban

League's annual report, in which their president declared, "[r]acism is the pandemic within the pandemic."[15] A recent live Zoom installation by artist John Sims called "2020: (Di) Visions of America" amplified the connections between the overlapping sirens of America's simultaneous pandemics, by using the sound to connect stories attesting to the inequitable death toll of Covid-19 borne by Black and Brown Americans with those witnessing the inordinate rates of police murder and mass incarceration visited on that same population.[16] At one of the key transitions of the piece, the artist drowns out the voice of Dr. Lisa Merrit—she has been reading a letter written to a Black, Indigenous, Person of Color (BIPOC) patient who has died from Covid-19—with a siren that only grows more insistent, bleeding indistinguishably into the soundscape of the next sequence, in which Sims himself reads a letter written "after the merciless death of George Floyd" that begins: "Dear Police, I've been meaning to write you. . ."[17]

The overlapping sirens of 2020's crises and their resonant amplification of multiple interlocking racialized traumas inform my approach to an understudied 1943 short story on the relationship between listening, subjectivity racism, death, and the state by journalist and writer Ann Petry, also with a poetic and ominous title: "On Saturday the Siren Sounds at Noon."[18] If critics have not yet figured out where to place this story by Petry about the impact of New York City's air raid siren testing program on a Black defense worker nervously awaiting the elevated train, this chapter argues it is a story that we need, right here right now.[19] Petry's story exemplifies Paula M. L. Moya's finding that "literature can show how past oppressive structures do damage to possible future selves," a conversation of necessary urgency in our current moment. While the coronavirus is novel, the racism that has intensified it in the United States most certainly is not, and Petry's "On Saturday" enables a deeper understanding of how the social production of sound stratifies and racializes space, and even more acutely during a national emergency. By focalizing the listening experience of New York's air raid siren test through the auditory perspective of her unnamed protagonist, "a Negro in faded blue overalls," Petry deftly amplifies the resonance between the siren's material, sociopolitical, mythic, *and* cultural dimensions, particularly the role of the sonic color line in enforcing state power during emergencies and announcing differential access to the rights and privileges of citizenship for Black men and women. As Claudia Rankine observed in *Just Us* (2020), "space itself is one of the understood privileges of whiteness."[20] I examine here how that violent and shifting territorialization occurs by racializing and weaponizing sound, a process of segregation termed the sonic color line. Examining the divergent reactions of listeners to sounds of state authority such as sirens show how white supremacist power structures claim public space and state protection for white people.

In this chapter, I argue that the siren of Petry's "On Saturday" represents an especially rich site for what Moya calls a "socioformal close reading," an intensive re-reading attuned to how Petry uses formal and thematic textual features such as aural imagery to "mediate the historically situated cultural and political tensions" expressed through her work, like the sonic color line.[21] This chapter works at the intersection of sound studies and African American literary studies to follow the siren's vibrations through the different strata of interpretive frames that give it and "On Saturday" ever-deeper meaning: the literary, the historical, and the mythic. I begin by identifying the "stratified aural image" as a key element of Petry's writing style. Emily Lordi describes an artist's style as a signature voice, their way of "telling a new story and making the material text of that story look and sound distinct."[22] Through the overwhelming and enveloping central image of the siren, Petry's "On Saturday" examines the emotional and political power of sound and experiments with "stratified aural imagery," or extended representations of competing narratives through layered, simultaneous sounds. I place this story within the larger "trope of the listener" in African American literature, a signifying chain of communication across texts and swaths of time, where Black writers portray characters in situations where listening becomes their primary sensory experience and source of information.[23]

Then, I trace the specific geography of Petry's story as well as the historical context of the air raid siren tests that inspired it, revealing how the sonic color line structured segregated space in the 1940s and how the air raid tests—a largely forgotten sound embedded in New York City's past—worked to reinforce the racializing of space via sound. Petry's depictions of the sirens and their differential resonance in "white" and "Black" neighborhoods help us understand the divergent listening habits of white and Black people in regard to state-declared emergencies as signs of whose lives are valued, protected, and central to the citizenship ideal. I "un-air" these sounds from their historical strata in order to make audible the connections between the New York City neighborhoods declared "dead spots" during the Second World War—areas where residents could not hear the sirens—and the overwhelming amount of emergency sirens putting today's medical and political "hot spots" on blast during the intertwined pandemics of racist policing and Covid-19.

Finally, I close by exploring the stratum of myth and conjure in "On Saturday," arguing that Petry's realist sonic details also engage with and challenge the deeply embedded misogyny of Western culture circulating via resonance with Homer's iconic "sirens." Michael Bull's recent book *Sirens* opens with a declaration of the rarity of writers who thoughtfully examine the deep entanglements of historical sirens with their Greek mythical archetypes: "Few have ventured to join those sirens we hear in war and on our streets to those transformed from the written page to appearing on the musical stage of the world, in film, on the printed page and, of course, in social thought."[24] It is all the more important, then, to explore Ann Petry's 1943 venture, which links the sirens of war and street with violent cultural representations of Black women as loudly sirenic—hypersexual, uncontrollable, seductive, and destructive—a sharp edge of the sonic color line that distorts and silences them. Between her New England education and pharmacy degree, Petry undoubtedly knew Greek mythology (and Latin) well, and she mobilizes the symbols here to point to their utility for white supremacy.[25] The narrator of Petry's story is ultimately seduced and destroyed by the siren's song of white American patriarchal power, not his wife Lilly Belle as he first assumes. At the end of the story, Petry conjures the state's siren of protection into a moaning funeral dirge for this Black family, entrapped by racism and misogynoir, defined by Queer Black scholar and activist Moya Bailey as "the unique ways in which Black women are pathologized in popular culture."[26] This chapter enacts a richer listening to Petry's signature "stratified aural imagery" as well as to her literary legacy, honoring the craft of the writer who brought "On Saturday" 's siren crying and wailing to life and encouraging readers to listen ever more deeply to its echoes: past, present, and future.

## Siren Sounds

Petry gave the first short story published under her own name a rhythmic, sibilant title that cast the mundane regularity of the New York City air raid tests in the vein of Lovecraftian horror: "On Saturday the Siren Sounds at Noon." Such a move ironizes the overt racism of white horror writers like Lovecraft and anticipates the blurred aesthetic of contemporary Black horror such as Jordan Peele's *Get Out* (2017) and the 2020 HBO series *Lovecraft Country* that captivated American audiences during Covid-19 quarantines[27]—and films that combine realist representations of racial trauma with supernatural imagery to represent how "Black history is Black horror," as horror writer and scholar Tananarive Due states.[28] In a 1947 interview, Petry described the inspiration for "On Saturday" as a blending of her experience of the material and psychological power of the air raid siren with a horrible Harlem fire she reported for the *People's Voice*:

> One Saturday I was standing on the 125th Street platform of the IRT subway when a siren suddenly went off. The screaming blast seemed to vibrate inside people. For the siren seemed to be just above the station. I immediately noticed the reactions of the people on the platform. They were interesting, especially the frantic knitting of a woman seated on a nearby bench . . . I began wondering how this unearthly howl would affect a criminal, a man hunted by the police. That was the first incident. The second was a tragedy I covered for my paper. There was a fire in Harlem in which two children had been burnt to death. Their parents were at work and the children were alone. I imagined their reactions when they returned home that night. I knew also that many Harlem parents, like Lilly Belle in the story, often left their children home alone while at work. Imaginatively combining the two incidents gave me my story.[29]

Tying together both incidents, Petry's "On Saturday" represents the horror lurking within the commonplace, or, more accurately, the feeling of living through the horrific *as* commonplace. Testing an air raid siren every week on Saturday attempts to contain and banish fear of an unknown attack, yet its "unearthly howl" also reminds listeners of the ever-present possibility of death. For some, as in the case of "a man hunted by police," the siren warns death lurks ever closer. The second incident was one of several deaths Petry covered as a journalist. Unfortunately, "she had found in Harlem extraordinary rates of domestic violence, alcoholism, rape, and murder" that prompted her to "ben[d] her fiction to explain the statistics that regularly confused people and seemed to support the hoary notions of biological inferiority."[30] Petry heard the police and fire sirens sounding so much more frequently in Harlem as a signal that racism was devastating American society, yet these sounds were explained away by white supremacist ideology as a signal of the need for racial segregation, normalizing the everyday expendability of Black lives in a white supremacist society.

Petry first alerts readers to the complexity of the image of the siren in the tagline accompanying the story in *The Crisis*. It reads, "to the average citizen an air raid warning is just another nuisance, but to this worker it brings memories of marital unhappiness and tragic love."[31] This disclaimer alerts readers to be attentive to the layered nature of the sound of the air raid siren; while official discourse attempted to fix the meaning of this sound for the "average citizen" in a nationalist matrix of wartime preparation and readiness on the domestic front, it was also considered a weekly annoyance by many and a horrifying reminder of "tragic love" to Petry's lone protagonist.[32] While this preview posits the worker's reading of the sound as aberrant—the syntax of the sentence bars him from the category "average citizen"—it also issues a challenge to readers to listen beyond the surface throughout the story.

"On Saturday" represents via sound the extreme vulnerability of the lives of Harlem's Black working poor, whom white oppression regularly pushed to the brink of existence. The story opens at five minutes to noon on a late-winter Saturday at the 241st Street elevated train station in the Bronx. Although periodization based on European literary movements has long placed Petry's writing as "realism" or "naturalism," Black New Yorkers would have recognized "On Saturday" as a horror story from the opening lines that find an unnamed "Negro Man" alone in a sea of white Bronx residents on a subway platform in a predominately white middle-class neighborhood named after the plantation where America's first president spent his childhood learning to become an enslaver.[33] Petry's setting already triggers internal warning sirens, especially when readers learn that he's on his way to work and has swerved from his usual routine to "get a breath of fresh air." The 1939 *WPA Guide to New York City* describes the 241st station as a springboard to nature, indicating its spatial and figurative distance from Manhattan's dense concrete metropolis, where Harlem is located: "[F]rom Spring through Fall, groups of hikers mill about the East 241st Street terminal of the IRT White Plains subway, for here begins a two and a half mile trail to sylvan Tibbet's Brook park in Yonkers."[34] Petry's detailed geographic pinpoints enable readers to map key locations, stylistically constructing a sense of realism while leaving a palpable historical record of the terror produced through the racialization of space in the Second World War–era New York City, and how Black people could—and couldn't—exist within and move between neighborhoods. It makes absolute sense that a Black man in "worn overalls" who lives at 219 133rd Street in Harlem would pace apprehensively by himself in Wakefield, deliberately keeping clear of the unfamiliar white people gathered at the other end of the platform—therein lies trouble.

That Petry leaves her protagonist unnamed is ambiguous and multivalent. On the one hand, his anonymity makes him a stand in for "everyman," particularly as the overalls, metal lunchbox, and Saturday swing shift suggest he is a defense plant worker, a citizen doing his part for the war effort. On the other, identifying him as a nameless "Negro man" signals his interpellation into the US social order via the veil of race, stripping him of his individuality and marking him with white people's fears, fantasies, and stereotypes, even as Petry carefully marks him as singular: "a Negro" rather than "the Negro." This identifier is also reminiscent of the tone of a news story, foreshadowing tragedy, while at the same time suggesting that the main character may want to remain anonymous, that he is on the run or in hiding.[35] However, while Petry never gives the protagonist's name, she focalizes her narration intimately through his stream-of-consciousness. In the opening paragraphs of the story, his thoughts move rapidly from present to past based on his visual observations; the shiny gleam of the train track pulls him back to a memory of working as a porter at a white hotel for example, and all the metal he had to shine, including a spittoon.[36] He keeps trying to shake off the memories that arise, but everything he sees, even posted advertisements, pull him back painfully to prior moments. And then, right on time, the air raid siren goes off for its weekly test. Even though the siren goes off as scheduled, the protagonist's reflex response is no less fear-driven. He "jumped nearly a foot when it first sounded."[37] While everyone else on the platform seems to ignore or endure it, the man visibly jumps, doubles over, and jerks around in agony at the loud, haunting sound that "was making a pounding pressure against his chest."[38] Echoing other sirens in the man's memory—"cops and ambulances and fire trucks"—the loud vibrations attack and completely take over the man's consciousness and body.[39] The siren's plaintive wail sparks a battle between his conscious mind—determined to suppress the agonizing memories of the death of one of his children in a tenement fire and the murder of his wife by his own hand—and his subconscious mind, which resurrects and reanimates these tortured moments until he throws himself in the path of the approaching train.[40] The blaring siren reverberates through the protagonist's mind and body, exacting his confession to the murder and ultimately accompanying his suicide, living on even after the train stops, leaving "a thin echo of the siren in the air."[41] The siren in "On Saturday," a character in its own right, functions as a stratified aural image revealing new layers of terror and meaning as its vibrations burrow ever deeply into the protagonist's life experiences, each layer of horror amplified by the one before it.

## "It Started as a Low, Weird Moan": The Siren's Stratified Aural Imagery

One of the most significant literary strategies that Petry develops in her fiction is her use of aural imagery, or written descriptions that both represent sound and transmit the experience and sensation of hearing. While Petry's aural imagery encompasses many varieties of sonic prompts to activate a reader's inner hearing—dialogue, music, screams, laughter, and ambient soundscapes—one of her unique stylistic signatures involves what I identify as "stratified aural imagery": the strategic layering of multiple simultaneous sounds atop one another that compete for coherence and dominance within the context of a narrative and in the comprehension of its readers.[42] "Stratified" is doing a lot of work for me here as a descriptor, just as Petry's representations of layered sounds perform multiple simultaneous creative, critical, and philosophical labors in her writing. "Stratified" is of English-language origin, a term developed during what Europeans consider the Enlightenment period, first used by eighteenth-century geologists and archaeologists to reference the new ways in which they perceived the earth as being constituted of layers of various substances—topsoil, detritus, rock, molten lava, and so on—that changed as one dug down deeper toward the planet's core. In biology, the term was taken up to refer to the composition of various living cells and to agricultural practices that layered various components of soil to promote better germination. In chemistry, the term referred to the ways in which various gasses separated and arranged themselves, particularly in the atmosphere. Just like "evolution," racial scientists, journalists, and academics put "stratification" to use in the social Darwinist period of the nineteenth century, to describe the development of social, racial, gendered, and economic hierarchies.[43]

"Stratified" then references several things at once in Petry's fiction. First, the way she exploits the properties of hearing as a sensory modality—that humans hear through multiple channels at once on what Ralph Ellison would later call the "lower frequencies" in *Invisible Man*[44]—picking up resonant vibrations through the skin, soles of the feet, and bones while taking in several simultaneous sounds as waves through the ear's auditory canal. "Stratified" also references how the brain processes these simultaneous sounds, using cultural and historical filters to determine which sounds should be ranked most important and listened to consciously on down to which sounds should be tuned out and ignored as much as possible even as their waves still touch one's body and impact feelings, thoughts, actions, and impulses nonetheless. Finally, I use "stratified" to reference the spatial orientation of Petry's representations of sound in her fiction, how she stretches a sound out in time—or, more accurately, how she stretches time out via sound, refusing linear narratives with flashbacks and jumps forward to create the sensation of downward movement through strata, even as her characters try to move upward literally for "fresh air" and metaphorically to better living circumstances.[45]

## "That Old Air Raid Alarm": Flopperoos, Beastly Wails, and Dead Spots

The first stratum of "On Saturday" I examine is historical: the blaring yet mundane weekly air raid siren test first heard in New York City in 1941. It eventually took over five months in the aftermath of Japan's airstrike on Pearl Harbor's US naval base to build anything resembling an all-city warning system operational on a mass scale. Just days after the bombing and the United States's official entry into both the European and Pacific theaters of the war and feeling themselves newly susceptible to a German airstrike, the city government designated $25,000 toward building an audio warning system to alert residents of an impending bombing, theoretically giving them time to find cover in below-ground shelters. The *New York Times* had been covering European air raid siren systems for months, particularly in reportage of Germany's blitz of London, but until December 1941, most New Yorkers had only vicariously imagined their "shrill shriek."⁴⁶

For New Yorkers, the official start of the war—and their newfound vulnerability to it—was signaled by loud, strange, and often terrifying sounds that reverberated in the city's airspace as company after company struggled to gain the civilians confidence and the fat city contract. Just two days after Pearl Harbor, the city tacked together a system composed primarily of police and fire sirens, alerts already laden with racialized meaning.⁴⁷ At least one person tried to keep American cities from wartime soundscapes permeated with the "beastly wail" of the London alarms: Dr. Winifred Fluck, from the University of Wales, who suggested to defense workers in Newark, New Jersey, that "maybe you can find a sort of 'victory chimes' which will have an 'up and at 'em sound' and not this scary, frightening alarm."⁴⁸ The *Times* reported Dr. Fluck's aggressively cheerful suggestion with neither irony nor sarcasm, but all the city tests that followed offered some echo of Britain's "beastly wails." Fog horns with a "whip-like" blast sounding something "like the repeated cries of a bird at night" and police-car mounted silo-drones were both tested on December 16, 1941.⁴⁹ They were, however, deemed "a flopperoo," as one observer told the *Times*. Both tests left wide swaths of the city "spared even a faint whine," including Greenwich Village, the Grand Concourse, Staten Island (then called Richmond), the Bronx, Washington Heights, and Harlem: all areas where Black and Puerto Rican people and newly arrived immigrants made their communities in the city.⁵⁰ The accompanying image from the story depicts an all-white, all-male group of workmen from the Consolidated Edison plant in Murray Hill, Manhattan, smiling and plugging their ears.⁵¹ Despite the headline that "New Air Raid Siren Does Best Job Yet," a round of tests in March of 1942 involving a giant siren made by Bell Telephone Laboratories that sounded "piercing wails" from the Manhattan Bridge revealed it struggled to be heard above an elevated train on Third Avenue or the rumbling traffic on West Street in the West Village.⁵² After the failure of the siren to be heard in many places in Manhattan, south of Canal Street—including New York's political and financial centers City Hall and Wall Street, to this day, one of the city's least residential areas—there was a special re-test just for this site. City Council Person Joseph Kinsley, who for years had represented the South Bronx in District 8, where three quarters of the Bronx's Black population lived by 1940 and almost all of its Puerto Rican residents, charged Mayor LaGuardia with issuing conflicting directions, suppressing a report on the best technical research on air raid alarms, and only equipping the city with "but two or three air raid sirens of questionable value, located in the most remote precincts of the city."⁵³ Whether or not New York City would ever be attacked by air, the air raid alerts became audible signals of whose lives were valued in New York City and who were most deserving of the city resources.

New York City Police Chief Valentine used the acoustic term "dead spots" to describe places where the sound of the siren couldn't be heard, which took on a frightening connotation when the sounds themselves were supposed to warn residents of incoming aerial bombs. In one of the tests in late December 1941, *The New York Amsterdam News* reported Harlem residents "listening in vain for raid siren": "Harlem listened along with the rest of the city until it could listen no more . . . long after the rehearsal was finished, no one could be found uptown who had heard the alarm."[54] Furthermore, 10 percent of the city's public schools hadn't heard the sirens in previous trial runs, "part of this due to mechanical lapses and the rest to personal error."[55] The week before the long-awaited first sustained test of the full network of 390 sirens sounding on Saturday, May 30, 1942, at noon, LaGuardia attempted to allay residents' fears of the dead spots that would undoubtedly be disclosed. In early July 1942, New York finally reduced its thirty-three "dead spots" to twelve. Notably, however, their distribution overlapped with the poorer, most racially segregated and densely populated areas of the city: four in Manhattan, three in the Bronx, two in Brooklyn, one in Staten Island.[56] Even though the *Times* reported no dead spots in Queens, the city's leading Black newspaper, *The New York Amsterdam News*, was still sending out "sound scouts" to check on the siren's coverage, reporting in August 1942 that residents of two Queens neighborhoods (Glendale and Ridgewood) "didn't hear a 'peep' " out of the latest super-siren placed in Brooklyn.[57] In the early 1940s, these neighborhoods hosted a large population of recently arrived immigrants from Eastern and Southern Europe whose access to the powers and privileges of homegrown American whiteness still remained tenuous during this period: Germany, Hungary, Italy, and Slovenia. The United States now faced these nations in war.

In a city already strafed with interlocking and spatialized systemic inequities of racism and classism, however, Black and Brown New Yorkers had long faced the grim realization that they lived in "deaf spots"[58] where the city was concerned— particularly during the lean years of the Great Depression—but now the reality of their neighborhoods as "dead spots" became even more frighteningly literal. In *Sirens*, Bull calls these alert systems "sonic instruments of the state, reinscribing the soundscapes of cities—placed throughout the city—high up on roofs of buildings or on lamp posts so that all can ideally hear their subject populations become part of a 'collective' in their juggling of fear, hope, and training."[59] In theory, then, air raid sirens are designed to be inclusive via the production of volume and the expertise of acoustical engineers. What happens, however, when the state-orchestrated testing and training and political battles over funding and the tactical worthiness of neighborhoods conflict with the imagined "collective" reception of the sound? The ideal of performative listening of a citizenship drilled into being through a uniform reaction to a sound signifying the approach of an unequivocal enemy—what Bull calls "an ideology of sonic hope"—is revealed as a fiction "laid bare at the very time in which states were proclaiming it."[60] This fiction is what Petry's short story takes to task along with the realities of the sonic color line.

The sound of the sirens—or lack thereof—spatialized hierarchies of race and class across New York City, while the sonic color line mobilized the ideologies of race and class to make gross assumptions about how New York's "great polyglot population" would listen to the sirens. A *New York Times* article published three days after the attack on Pearl Harbor and the day after the city's "first air raid warnings in its history" described city residents overall as "glassily indifferent," unwilling to leave their lunches in Times Square or their places in line at Fifth Avenue Christmas sales. One crowded elevated train car collectively refused to evacuate, staring down an Air Raid Warden until he left in a huff.[61] Although the piece initially describes the apathy as collective, the sonic color line guided their speculation when the *Times* writer came to characterize the reaction of different New York City neighborhoods. Whiter neighborhoods clamored for louder sirens and remained "tense" and "alert"; Brooklyn was especially "serious," with families having "emotional reunions" with their school children and "clustering around radios," an image I have discussed as racialized in *The Sonic Color Line*.[62] The largely immigrant Jewish, Irish, and Eastern European East Side—a stronghold of Socialism and labor activity throughout the 1930s—was simultaneously "indifferent and confused." Both Little Italy and Chinatown were ethnically and racially stereotyped as "voluble" and "jabbering," respectively.[63] Finally, rather than investigate whether or not the system was heard in Harlem—as the *Amsterdam News* reported, it was not—the *Times* instead characterized New York City's Black and Brown residents as "nonchalant" toward the danger heralded by the air raid siren and the drive to unify the United States in the war effort.[64]

However, in "On Saturday," Petry doesn't represent Harlem as "nonchalant," but distant from the siren's auditory footprint. Petry's protagonist, a Harlem resident, compares the startling loudness of the siren in this woodsy area of 241st Street with his neighborhood in Harlem near 133rd Street, where ever-present police sirens and other "street noises dulled the sound of its wail." Often he was "underground in the subway when it sounded," headed out to work on what is a day of leisure for the middle class.[65] The stratified aural image of Harlem's noise drowning out the air raid warning evokes the neighborhood's marginality while suggesting that the imminent danger for Harlem's residents was domestic rather than foreign. The daily noise of US racism, grinding poverty, and servitude dulled the siren's triumphant proclamation that the war to end all racism and fascism was being fought abroad.

In the noisy gear-up to US involvement in global warfare, Black and Brown residents of de facto racially segregated New York questioned the notion that "the enemy" only lived across oceans and quite literally struggled every Saturday at noon to hear any semblance of protection from the state. In a January 1942 reader letter to the *Pittsburgh Courier*, one of the leading Black newspapers in the United States, 26-year-old James G. Thompson questioned the sudden push for Black men to fight for a nation that had never ceased to violently exclude them—"should I sacrifice my life to live half American?"—and inspired the *Courier*'s "Double V Campaign"—by asserting that people of color faced domestic enemies whose racism was just as dangerous as Germany's: "Let we colored Americans adopt the double VV for a double victory. The first V for victory over our enemies from without, the second V for victory over our enemies from within. For surely those who perpetrate these ugly prejudices here are seeking to destroy our democratic form of government just as surely as the Axis forces."[66] In many ways, the weekly two-minute air raid siren tests sounding in New York until shortly after V-E ("Victory in Europe") Day in 1945 provided stunningly audible evidence of the need for the Double VV campaign. The testing essentially sounded out how New York City and cities across the country had spatialized racism in the twentieth century, particularly after Black Americans migrated to the urban North in great numbers around the onset of the First World War.

The sounding of the air raid siren had the power to invisibly express the racial order of things in the white supremacist society of the 1940s. Regardless of its actual efficacy in saving lives, to live within earshot of the siren's wail provided reverberant feedback that one was a valued and protected citizen of the city and the nation, worthy of being saved from impending death. Ruth Wilson Gilmore defines racism as the "state-sanctioned or extralegal production and exploitation of group-differentiated vulnerability to premature death."[67] This affect is key to what Bull calls the state's "ideology of sonic protection," which is deeply entangled in the sonic color line.[68] Living within the range of the siren's wail meant an assumption of protection and citizenship privilege; one was considered a valuable part of the nation in the event of danger. To live outside of this threshold, or, more accurately, to live where state authorities chose not to extend the air raid signal—whether due to passive unconcern, strategic prioritizing, aggressive suppression, or likely a mix of all three sentiments—meant existing with the constant, unequivocal knowledge that you were inessential, not worth protecting or even being given a fighting chance of escaping death. To hear silence when the sirens were supposed to sound meant that one's life, ultimately, did not matter. When Petry's protagonist first hears the siren, he dismisses it "contemptuously" as "that old air raid siren," yet he realizes it "made him uneasy" nonetheless.[69]

Fittingly, Petry described the subjects of her fiction in terms of war imagery: "the walking wounded," a triage term that describes injured people of the lowest priority. According to medical definitions, the walking wounded appear to be better off than others with more visible injuries; they are ambulatory and seem likely to survive while waiting for aid.[70] Petry's deft characterizations, however, tend to unseen wounds both deep and wide in Black people. Her fiction is attentive to the pain and grief that the white listening ear's stereotyping and fetishization of "resilience" have tuned out in order to continue extracting labor, wealth, and privilege.[71] Both the unnamed protagonist and his wife Lilly Belle are "walking wounded" in "On Saturday" and Petry un-airs their story through the invisible but heard and deeply felt material impact of the air raid siren. Before the siren begins its "low weird moan," the man is "on his way to work in the Bronx" as if it is any old Saturday and he had not endured immense trauma over the last few weeks and lashed out in murderous violence just that morning.

The surprising loudness of the siren overtakes the main character in the midst of his fear, anger, and pain, especially because he is unaccustomed to hearing it *at all* let alone as loudly as it sounds in white neighborhoods. The shock of such volume, especially of a sound reminiscent of law enforcement, breaks through his attempt to repress the violence he has suffered and committed in turn. The siren's vibration—a conspicuously audible but unseen force of control—begins to physically abuse him as he tries to cover his ears to no avail. An invisible but formidable adversary, the sound starts "making a pounding pressure in his chest . . . hitting him in the stomach" and appears to assault and possess him, turning his chest cavity into an echo chamber where his hijacked heart beats "faster and harder" to this new tempo and drags his mind back to the past that he was trying so desperately to escape.⁷² As Petry shifts the siren's imagery from a "higher screaming note" to a "low, louder blast," his internal wounds are such that he can no longer walk. The siren has violently "pinioned him where he was"; it "hit him all over until he doubled up again like a jack knife," causing him to drop his lunch pail.⁷³ Even if he wanted to, the protagonist is unable to move toward the people at the opposite end of the platform, who are not reacting to the siren in this way, or at all. They appear to ignore both the siren and the man, as "they look down on the street and soak up some of the winter sun" as if he doesn't exist, as if he is not in excruciating pain in front of their eyes, and, as if they all aren't supposed to take cover for the drill.⁷⁴ Because "white privilege comes with the expectation of protection and preferences, no matter where [one] lives in the country, what job [one has] or how much money [one] makes," they know that the police will not harass them, and that they are likely not susceptible to whatever the man's pain may be.⁷⁵

The sound of the siren also reveals that the Black man at the train station knows the converse is true for him, that because safety in the United States is a relative property of whiteness, being in a more "sylvan" part of the city actually puts him *more* at risk because the white people there consider his very presence *as* the danger. He always already possesses double consciousness, as W. E. B. Du Bois powerfully theorized; therefore he knows that, although subway stations are ostensibly public, he is stepping out into white segregated space, and he must be on his edge. He knows whites will perceive him as a racialized trespasser from a neighborhood they imagine as dangerous (for them) and they ensure as deleterious for Black people by crafting unequal social, political, and economic policies. It is little wonder, then, that the siren sounds much more loudly to the man's ears than it does to the others at the station. He's already apprehensively listening to and for the sound of white fear. As J. Martin Daughtry argues in *Listening to War*: "life in wartime . . . radically increases the salience of sound for those who are struggling to make sense of their surroundings."⁷⁶ Daughtry helps us hear the doubled sounds of warfare in this story. Yes, the testing of the air raid sirens reminds all listeners of their proximity to the war "over there" as well as the "swiftly increasing threat of destruction visited upon subject populations in times of war."⁷⁷ However, this man's visible embodied trauma upon hearing this siren serves to reminds readers that he is a combatant engaged in war right then and there on the platform, in New York City, in the United States, as well as his own neighborhood in Harlem—spaces that begin to bleed together the longer the siren wails, pulling him back to the fire that killed one of his children. Until there is Double Victory against racism, abroad and in its domestic theater, there are no truly safe spaces, and the futurity of his family remains threatened.

## "I Can't Even Hear My Own Voice": Silence, Echo, and the Siren's Song

Filtered as it is through the consciousness of a traumatized, angry husband who only acknowledges murdering his wife at story's close, "On Saturday" reveals how tragically and deceptively easy it is to think of Lilly Belle only as a femme fatale to blame for all of the family's death and destruction. However, I argue that Petry's stratified aural imagery encourages readers to suss out the dominant white supremacist narratives about Black women during this period—as good time girls, irresponsible mothers, devious Jezebels and/or resilient super women—noting how these representations converge to silence Black women while simultaneously blaming them for the circumstances racism, poverty, and misogynoir have wrought. Petry's protagonist is quick to assume Lilly Belle is wholly responsible for their child's death, an assumption that ultimately destroys them both while submerging the larger systemic and economic forces at work. This chapter's final section connects the literary, geographic, and historical layers of Petry's siren image with its symbolism at the level of myth and folk epistemologies, particularly in terms of intersectional inequality in the United States—how white supremacy's domestic warfare against Black people impacts intimate life between Black men and Black women.

The mythic stratum of Petry's air raid siren signifies on the sirens of Homeric legend as well as their echoes in the discourse of 1940s "screen sirens" and glamour girls. The name Lilly Belle evokes the "siren," because of the sonorous connotations of the homonym "bell" but also that lilies are traditional funeral flowers and "belle" translates to "beautiful" in French. Duke Ellington wrote the homage "The Blue Belles of Harlem" for Paul Whiteman's Orchestra in 1939. "[T]he title is self-explanatory," Ellington noted in his autobiography.[78] Dan Burley's "Back Door Stuff" column frequently discussed the belles of Harlem nightlife for *The New York Amsterdam News*. One such entry featured a letter from "Disillusioned Glamour Gal of 1940" warning young Harlem women to start "working for an honest living" rather than being "abhorrent to work" like her and her friends, who "sit like brownstone Grecian Sirens around some bar" waiting for a man with some money to come along.[79] Brownstone here is elegantly doubled imagery, referencing the women's brown skin and the stacked thickness of their forms as well as the specific type of housing in Harlem that these sirens are after, just as some readers may assume Petry's character Lilly Belle to be. The notion of these women as "abhorrent to work" tracks with the very first glimpse readers get of Lilly Belle through the narrator's eyes as he "frowned down at her" for being "still asleep" on a Monday morning and "untidy and bedraggled" while doing so. And as her appearance goes, so too, the frown implies, must go the house and the children.[80] Even though keeping Lilly Belle up in the apartment without a job outside of the home represents a status symbol in a patriarchal society—and the protagonist deeply cares for his children—heteronormative gender roles trigger a corrosive resentment for the protagonist, who leaves the house early and angry each morning to a job whose pay is stretched to the limit. While Petry doesn't make readers privy to Lilly Belle's thoughts, feelings, and wishes, her resistant actions suggest she, too, is made miserable by this aspirational—read: white middle class—domestic arrangement.

To challenge the blaring misogyny at the heart of such enduring Western cultural mythology, Petry's sirenic imagery engages Du Boisian concepts of double-consciousness, "second sight," and the veil—images themselves combining ancient Greek thought and Black diasporic practices as I have argued elsewhere—and anthropomorphizes the siren as the "moaning," "sobbing," "talking" voice at the center of the story. The siren shares knowledge that Solimar Otero calls an "archive of conjure."[81] In her study of Afro-Latinx spirit mediums and stories of the dead, Otero defines "archives of conjure" as "a set of spiritual, scholarly and artistic practices based on the awareness of the dead as active agents that work through imaginative principles."[82] Otero's study theorizes "sirens" in the context of stage performances and literary characterizations that call upon the hybrid, transgendered oceanic spirit Erinle, an "alluring and ambiguous" sirenic figure who appears in many forms in Afrolatinx Caribbean culture and can be traced back to Nigerian divinities. Stories of Erinle depict them as experiencing exploitation and abuse at the hands of the god Yemayá, who kidnapped them to the bottom of the ocean and held them there as a sexual prisoner. Yemayá released Erinle only upon the extinguishing of her desire, but not before cutting out their tongue so that Erinle couldn't share their knowledge about all that they had seen at the bottom of the sea.[83] Petry echoes the story of the siren Erinle and their captor Yemayá in the unequal and violent gender dynamics in "On Saturday," specifically Lilly Belle's feelings of entrapment and her husband's anger at her behavior and at his now-lapsed desire for her. Yemayá's cutting out of the tongue resonated in the protagonist's murder of Lilly Belle and the way in which the siren's shifting extraverbal sounds communicate her pain *and* knowledge.

Petry's knowledge of Black diasporic folk practice such as the story of Erinle and Yemayá has always been underestimated. Petry scholar Kenneth Clark notes that she always had a deep-seated interest in and familial knowledge of a tradition he notes is "too often solely associated with black Southern writers."[84] Petry, a pharmacist like her father and aunt, traced the family business back to the stories of her paternal great-great-aunt Hal, who had been a conjure woman in Hartford, Connecticut, and, though she worked for, as Petry described, "a perfectly legitimate doctor of medicine," she "did most of the prescribing" with her roots and herbs. In a 1988 autobiographical essay, Petry expressed how she identified with her family's experiences and the knowledge they passed down through story, making it known that she *is* her ancestors and announcing: "I am a conjure woman."[85] Petry's knowledge of the mythic archetypes and alternate cosmologies of conjure and the siren's symbolic connections to them encourage a re-reading of "On Saturday" attuned to Lilly Belle's silences and the siren's sound as an archive of conjure embodying her suppressed voice and restless spirit. The vibrations enter the man's body, force him to listen to her, accept his culpability in her murder, and receive the knowledge of the systemic violence that ensured the entire family was fatally "locked in" to their dark, narrow, and dangerous tenement apartment long before the fire.[86]

The narrator doesn't consciously come to understand the deep systemic causes of his precarious condition until the long, extended cry of the siren enters his body, possesses him, and recasts his memory, replaying it for him as if he is watching himself in a movie.[87] With the siren's "noise tearing inside him, he could see himself as clearly as though by some miracle he'd been transformed into another person."[88] The image of doubling itself has a double meaning. On the one hand, the narrator embodies the effects of Du Boisian double-consciousness, a "second-sight" that allows him to see the racial formation of the United States more clearly. The wailing moan of the siren functions as what Du Bois dubs the "revelation of the other world" in *The Souls of Black Folk*, and it divides him into two selves, reconfigures the protagonist's senses and perspective on the world.[89] It begins: "[H]e could see himself coming home on Monday afternoon. It was just about three o'clock . . . his lunchbox was empty and he could see it swinging light from his hand."[90] Du Bois's figuration of second sight draws from Black folk epistemologies, particularly that of conjure. Petry's stratified image of the siren also pulls from that diasporic tradition, operating as a kind of spirit possession in addition to a form of psychic clarity for the protagonist.[91] At one point, the protagonist notices that the siren overtakes his will and hijacks ability to speak: "[H]e moaned, 'make it stop.' And then in alarm, 'I can't even hear my own voice. My 'voice is gone.'"[92] Thus quieted and fixed in place, the siren's vibrations then put him into a trancelike state of memory that forces him to listen deeply through its layers until he relives the events of the fire and the murder, but from a different perspective, one outside of his own body. Petry calls attention to this doubling by repeating the phrase "he could see himself" over nine times as the protagonist works his way through the trauma of his vivid and visceral memories, forced and guided by the stratified sound of the siren's scream all the way back to the fire that destroyed his family's home, injured his children and caused the death of his daughter Eloise, as well as the fact that weeks later—just that very Saturday morning—he strangled Lilly Belle to death with her own mourning veil. He sees himself blaming her for everything after overhearing Cora, the building's janitor, confronting Lilly Belle about locking the children alone in the apartment so she could meet her lover.[93]

While the protagonist acts with certainty that Lilly Belle has been unfaithful and negligent, the memory that the siren pulls from him during the flashback sequence is much more ambiguous, revealing that he has repeatedly silenced Lilly Belle and refused to listen to her. In the few places that Petry represents her speech, it trails off and ends in dashes suggesting she has been cut off while speaking, such as the morning of the fire, after her husband threatens to kill her if she ever leaves their kids home alone again, she responds with "I'm goin' to have me some fun—."[94] When he remembers returning from the morgue to the hospital ward and seeing Lilly Belle dressed in a black mourning veil, he was thinking so intensely about her clothes that "he only half heard her" as she recounted her side of the story, only able to quote snippets to the reader that trail off: "'But I ran into Alice—and when I came back,' she licked her lips as though they were suddenly dry."[95] His bitterness and anger toward Lilly Belle make him unwilling or unable to hear anything she says, a tuning out that has dire consequences for both of them. The rage that drives him to strangle Lilly Belle comes from an overheard and obstructed conversation, one that he never actually confronts her about. The narrator hears only one side, the janitor's accusation that Lilly has been "never been no damn good" and her threat that if Lilly Belle doesn't "quit runnin' to that bar with that dressed up monkey, and stayin' away from here all day long, [she's] gonna tell that poor fool [she's] married to."[96] When Cora finished, "Lilly Belle said something that he couldn't hear" in response. It is especially interesting that Petry describes the conversation coming "clear to his ears like a victrola record or the radio," considering how she depicts the unreliability of radio as a conveyor of truth in *The Street*, particularly for and about Black women.⁹⁷ The victrola record references one of the only arenas where Black women could express love, sexual pleasure, and a yearning for freedoms outside of domestic expectations: the blues. In *Blues Legacies and Black Feminism*, Angela Y. Davis notes the "sparsity of allusions to marriage and domesticity in women's blues" and observes that "the protagonists in women's blues are seldom wives and almost never mothers."⁹⁸ It was also "one of the few cultural spaces in which a public discourse on male violence had been previously established" by Black women.⁹⁹ The ears through which Lilly Belle's husband hears her then are triggered by siren's songs—radio caricatures and the blues—and even though he remembers that "Lilly Belle said something he couldn't hear" in response to Cora's accusations, he fills in her side of the story himself with his assumptions and "very quietly" decides to grab her black veil and strangle her with it. As he recalls tightening the veil around her throat, the siren increases its vibrational pressure inside of him, until "his lungs were moaning and shrieking with agony," until he too feels as if he cannot breathe.¹⁰⁰

Lilly Belle's "long black veil that billowed about her as she moved" is an especially significant image of opacity at the climax of "On Saturday," a rejoinder to Du Bois's famous veil imagery in *The Souls of Black Folk*. Du Bois's metaphoric "veil" draws on a Black Diasporic belief that infants who are born with a caul still attached, a portion of the birth membrane remaining on the head, are said to be "behind the veil" and gifted with preternatural abilities, or "second sight." Du Bois's "veil" is a metaphor for racial blindness caused by the color-line, a figurative panel that the white power structure places upon Black people to obscure their humanity and render them socially, economically, and politically invisible. As whites cannot see through the veil of "blackness" they have constructed through fear, fantasy, and stereotype, this willfully distorted vision enables them to continually dehumanize Black people, characterizing them as abstract "problems" rather than individual human beings. However, the subsequent—and double-edged—consolation of solitary confinement within the veil is a striking clarity of insight into the artifice of the color-line and the empty but vicious truth of the concept of "race." Although it is "almost a burden beyond the measure of their strength," this "second-sight" equips Black people with incisive powers of social critique to deconstruct the workings of the color-line in American society.¹⁰¹ Petry's use of the veil in "On Saturday" suggests that Du Bois's schema does not account for gender as a field of power operative within the category of race. Within the veil— and entangled with the sonic color line—there are gendered forms of mishearing, miscommunication, and violence that distort intimate relationships.

After all, readers never hear the details of Lilly Belle's story, only the siren's cries and sobs. We don't know how she came to be married to the protagonist, or if she ever wanted to be married at all. We don't know whether or not she loved men or women or both, whether or not they were ever in love, whether or not she got pregnant before going to the judge or the preacher, or if they just took up living together. Did she ever consider her husband handsome or strong or trustworthy or just her lot in life? We do not know if she had family looking out for her when they met, or if she was fending for herself, if they met in a club or on the subway or through a friend. What we do know is "the white world had made the rules about how to be a man, how to be a woman, how to live intimately"[102] and these racialized norms punished Black women severely for wanting anything better for themselves or even just something different. Even though we know so little about Lilly Belle, is it difficult to imagine that she would "grow weary of being a Negro woman in all of the ways expected of her" by the white and Black worlds?[103] Do these questions enable an understanding of Lilly Belle not solely as an antagonist, but as a complex subject in her own right, one who might have been married to her husband but *not* to the sociopolitically imposed ideal of white middle-class patriarchal domesticity?

The story ends with the main character listening to Lilly Belle at long last, in the guise of the air raid siren. Whether or not his guilty conscience conjures up her sirenic presence or her sirenic presence conjures up his guilt, the way in which he heard the siren's wail enables him to realize that she was an easy, visible target for his grief and anger, but far from its root cause. Killing his wife for the damage created by her unhappiness, mistakes, and grief brought him neither consolation nor liberation, and it certainly did not bring back Eloise. As the man remembers the "cold hard knot that formed inside him when he saw that she was dead,"[104] he realizes that he has lost his humanity and entangled himself further in white systems of oppression, that the air raid siren portends police sirens and imprisonment. Upon this realization, Petry represents the siren as howling with renewed vigor, as the train approached "the siren took on a new note—a louder sharper sobbing sound. It was talking. 'Locked in. They were locked in.' " The repetition of "locked in" throughout this section onomatopoeically suggests the sound of the coming train while reminding the man that the "invisible walls of oppression locking his family into tenement life are manifested in the locked door preventing his children's escape."[105] Unable to change the past and feeling as if he will never be able to escape the present's radioactive insight, the protagonist sees no future for himself and throws himself into the path of the approaching train. The siren now sounds an elegy for Eloise, Lilly Belle, the protagonist, and the now-orphaned children still "locked in" to segregated Harlem. The images of the echo and the stratified sound of the siren in "On Saturday" suggest the devastating materiality of invisible structures of oppression during a wartime emergency, particularly the interlocking processes of racialization and gendering that only grew more acute. While the ending of this story offers no solutions, the last line hints at the necessity of remembrance, describing how "even after the train stopped, there was a thin echo of the siren on the air."[106] Here Petry evokes the myth of Echo, a woman whose power to speak for herself was revoked by the gods, dooming her disembodied voice to repeat the words of her lost love Narcissus long after her physical form had wasted away. It's a "potentially trapping sound"[107] of trauma as Tsitsi Jaji explains: Locked in. Locked in. Locked in. Interweaving "echo" with the multifaceted song of the siren suggests the simultaneous presence and absence of Lilly Belle's voice at the heart of "On Saturday the Siren Sounds at Noon" and of Black women's voices in the United States, a particular kind of intersectional violence of the sonic color line. However, the echo is not without its own powers. In *Africa in Stereo*, Jaji claims the echo as a "resonant trope" for articulating Black diasporic relations across multiple times and spaces, a mode of receiving and hearing connections between "histories that may be muted, or diminishing, but never fully silenced."[108] Carter Mathes notes in *Imagine the Sound* that the circularity of the echo can "disrupt narrative structures based on uncomplicated ideas of progress" like those of Western capitalism, while vibrationally "reducing the time and space of separation" between past presence and present absence.[109] Diminished yet present, trapped yet traveling, that "thin echo of the siren on the air" at the end of "On Saturday" remains, sounding out powerful and disruptive connections with the catastrophic events of 2020.

## Conclusion: Hot Spots

Moving back through the strata of Petry's siren imagery in February 2021, listening for contemporary echoes of disruption amplifies the deeply intersectional entanglements of the Covid-19 crisis. There are echoes of pandemics within pandemics within pandemics that bring the past and the present so palpably close as to chafe. In fact, a recent study shows, quite literally, that the same "dead spots" during the 1940s air raid drills are now today's Covid-19 "hot spots," the hardest hit neighborhoods where ambulance sirens wailed in intense waves throughout 2020.[110] Furthermore, America's deeply sedimented misogynoir—expressed in "On Saturday" by the overwhelming hyperaudibility created by sirenic myths about Black women coupled with the masking and muting of their voices until they must scream to even be noticed—echoes in Dr. Susan Moore's Facebook documentation of Indiana University hospital's mistreatment while she was hospitalized for Covid-19 in December 2020.[111] She died five days after discharge. The way in which Petry's story connects the political with the personal doesn't just document the connections between the persistence of racialized and gendered stereotypes and the corporeal violence of the fiction of colorblindness mobilized by the state during global crises, it embodies them—how they sound, look, feel. Petry's use of stratified aural imagery as a literary technique shows, yet again, that sirens and echoes of segregation have *never* been abstract or theoretical for those living this experience. The materialities of intersectional oppression resonate loudly and deeply in people's lives and with an intimate force that cannot be tuned out or ignored. And just as in Petry's moment, those for whom the war sirens rang loudest—people of color drafted for the most dangerous details, folks on the home front facing racist violence and extreme poverty while fulfilling essential factory jobs, women for whom no childcare support system existed—receive the least care and concern from a state claiming that it was doing its best to protect everyone and that "we" were all in this together. Ann Petry's "On Saturday the Siren Sounds at Noon" reminds us that the sonic color line isn't a static dictate or a fixed boundary; rather, it is something crafted anew with every refusal to listen, every act of explaining away and each expression of disbelief, private or public.

## Notes

1   I deeply thank Nathalie Aghoro for her amazing editorial work that strengthened this chapter immensely, as well as supportive colleagues at the Frederick Douglass

Institute at the University of Rochester, where I first presented this work, especially Jeff Tucker, Anthea Butler, Niambi Carter, and Millery Polyné. I extend heartfelt gratitude to Priscilla Peña Ovalle and John Lee for unfaltering support and conversation throughout the revision process.

2   These pieces, in the order they are quoted in this sentence, merely sample the vast discourse about the sound of sirens in New York City in 2020, all accessed on 19 January 2021. Doree Lewak, "New Yorkers are on Edge Over Relentless Sirens During Coronavirus Lockdowns," *New York Post*, April 1, 2020, https://nypost.com/2020/04/01/new-yorkers-on-edge-over-relentless-sirens-in-coronavirus-lockdown/; Julia Vitullio-Martin, "Sirens and Suffering: Rethinking the Soundtrack of the Coronavirus Crisis," *Gotham Gazette*, April 21, 2020, https://www.gothamgazette.com/opinion/9324-sirens-suffering-rethinking-soundtrack-coronavirus-crisis-new-york-city; David Remnick, "New York City in the Coronavirus Pandemic," *The New Yorker*, April 5, 2020, https://www.newyorker.com/magazine/2020/04/13/new-york-city-in-the-coronavirus-pandemic; Samer Kalaf, "Alone in the City of Sirens," *New Republic*, April 10, 2020, https://newrepublic.com/article/157256/coronavirus-new-york-city-sirens; Lindsey Zoladz, "Learning to Listen to, and Beyond, the Siren's Call," *The New York Times*, April 5, 2020, https://www.nytimes.com/2020/04/05/arts/music/coronavirus-sirens-music.html. There are also multiple recordings of the sirens uploaded to YouTube, from official outlets such as the Associated Press, "Birdsong, Blaring Sirens: A Pandemic in Sound," released April 6, 2020, video, 1:13, https://www.youtube.com/watch?v=UlNVa5qME64; and citizen journalists such as Sandi Bachom, "DAY28: Sirens and Refrigerated Trucks At NYC Hospitals 3/29/20," released March 30, 2020, video, 4:56, https://www.youtube.com/watch?v=4WmJ-f3SB4c.

3   "Governor Cuomo Signs the 'New York State on PAUSE' Executive Order," *New York State Government*, March 20, 2020, https://www.governor.ny.gov/news/governor-cuomo-signs-new-york-state-pause-executive-order; Quoctrung Bui and Emily Badger, "The Coronavirus Quieted City Noise: Listen to What's Left," *The New York Times*, May 22, 2020, https://www.nytimes.com/interactive/2020/05/22/upshot/coronavirus-quiet-city-noise.html.

4   Molly Jong-Fast, "Embracing Life in a Time of Death," *Vogue*, April 18, 2020, https://www.vogue.com/article/new-york-coronavirus-life-in-time-of-death.

5   Interactive map that, as of this writing, was frequently updated: "Coronavirus World Map: Tracking the Outbreak," *The New York Times*, January 19, 2021, https://www.nytimes.com/interactive/2020/world/coronavirus-maps.html.

6   "Amid Ongoing COVID-19 Pandemic, Governor Cuomo Announces New Hospital Network Central Coordinating Team," *New York State Government*, March 31, 2020, https://www.governor.ny.gov/news/amid-ongoing-covid-19-pandemic-governor-cuomo-announces-new-hospital-network-central.

7   "Stop LAPD Spying Coalition" qtd. in Pascal Emmer, Woods Ervin, Derecka Purnell, Andrea J. Ritchie, and Tiffany Wang, *Unmasked: Impacts of Pandemic Policing* (COVID19 Policing Project, 2020), 6.

8   "Hopeful Birdsong, Foreboding Sirens: A Pandemic in Sound," *Associated Press News*, April 11, 2020, https://apnews.com/article/d18d4fc6e2f4ff3f1d439fc12f6a3d98. There was also a flood of pieces about birdsong and New York in Spring 2020. For example, see: Antonio de Luca, Dave Taft, and Umi Syam, "New York Is Quiet: Listen to the Birds," *The New York Times*, May 31, 2020, https://www.nytimes.com

/interactive/2020/05/31/nyregion/coronavirus-birding-nyc.html; Erik Stokstad, "When COVID-19 Silenced Cities, Birdsong Recaptured Its Former Glory," *Sciencemag*, September 24, 2020, https://www.sciencemag.org/news/2020/09/when-covid-19-silenced-cities-birdsong-recaptured-its-former-glory; and Karissa Krenz, "How the Sound of New York City Has Changed during the COVID-19 Lockdown," *WQXR Editorial*, April 30, 2020, https://www.wqxr.org/story/how-sound-new-york-city-has-changed-during-covid-19-lockdown/.

9   Zoladz, "Learning to Listen to, and Beyond, the Siren's Call," n.p.
10  Amir Vera and Laura Ly, "White Woman Who Called Police on a Black Man Bird-Watching in Central Park Has Been Fired," *CNN*, May 26, 2020, https://www.cnn.com/2020/05/26/us/central-park-video-dog-video-african-american-trnd/index.html. For more on Christian Cooper, racism, and NYC birdsong, see Olivia Giovetti, "Notes on Birdsong: A Dual History of Racism and Nature's Music," *Van Magazine*, May 29, 2020, https://van-us.atavist.com/notes-on-birdsong.
11  Zoladz, "Learning to Listen to, and Beyond, the Siren's Call," n.p.
12  Ibid.
13  Kalaf, "Alone in the City of Sirens," n.p. Emphasis mine.
14  Ibid.
15  Mark Morial, qtd. in *State of Black America, Unmasked 2020: Executive Summary* (A National Urban League Publication, 2020), 6, http://sobadev.iamempowered.com/sites/soba.iamempowered.com/files/NUL-SOBA-2020-ES-web.pdf.
16  John Sims, *2020: (Di)Visions of America*, The Ringling Museum (Sarasota, Florida), January 18, 2021, virtual streaming program, https://www.ringling.org/2020-divisions-america-performance-john-sims.
17  Ibid.
18  Ann Petry, "On Saturday the Siren Sounds at Noon," *The Crisis* (December 1943): 368–9. This is the first short story published under her own name, but not her first published story, which is "Marie of the Cabin Club" (1939). This story, too, has sound as a central focus, particularly the trumpet and the sound of jazz.
19  Mentioned primarily in the contexts of true crime and melodrama, "On Saturday" has yet to receive much consideration (Hillary Holladay, *Ann Petry* [New York: Twayne Publishers, 1996]; Hazel Arnett Ervin and Hillary Holladay, eds. *Ann Petry's Short Fiction: Critical Essays* [Westport, CT: Praeger, 2004]). There are several reasons why this story has been referenced largely as an interesting footnote to the publication of *The Street* (1946). First, "On Saturday" is the first story Petry published under her own name, the story that gained her the attention of Houghton Mifflin editor Elenor Daniels, who read it and wrote to Petry in 1944 asking, "Are you by any chance working on a book?" and alerting Petry to the fellowship that she applied for—and won—the following year (Qtd. in Lawrence P. Jackson, *The Indignant Generation: A Narrative History of African American Writers and Critics, 1934-1960* [Princeton, NJ: Princeton University Press, 2011], 146.). Second, there's the issue of white gatekeeping and limited access, intertwined with the ongoing racism and sexism of academic literary study. Petry, as many scholars have noted as late as 2013, is herself critically understudied, in no small part to the narrow positioning of Richard Wright and Ralph Ellison as *the* main Black writers in the 1940s and 1950s, along with academic elitism that shuns women writers in particular for their commercial success: *The Street* sold over a million copies upon its release (Kenneth Clark, *The Radical Fiction of Ann Petry* [Baton Rouge:

Louisiana State University Press, 2013]). Furthermore, "On Saturday" was published in *The Crisis*, a Black press publication known for its political writing, not a white "mainstream" literary magazine, nor was it collected in Petry's 1971 anthology of short fiction *Miss Muriel and Other Stories*. Until mystery writer Paula L. Brooks reprinted it in her edited collection *Spooks, Spies, and Private Eyes: Black Mystery, Crime and Suspense Fiction of the 20th Century* (New York: Doubleday, 1995), "On Saturday" was only available via microfiche and archival copies of *The Crisis*. Brooks's collection contextualized Petry's little-known story within a longer history of Black mystery and crime writing stretching back through the late nineteenth century and into the contemporary fiction of Walter Mosley and Barbara Neely. So-called genre fiction—a term intended to devalue and set science fiction, horror, mystery, romance, and so on apart from "serious" literature—has also yet to be given its full due, although this is changing dramatically in our current moment. Even the term "horror literature" needs to be broken open beyond the limits of Western concepts of "story." In the introduction to *The Palgrave Handbook to Horror Literature,* Kevin Corstorphine discusses how it too often leaves out "folklore (oral traditions), religion, and mythology," all of which are central to "On Saturday." Corstorphine also discusses that part of the critical devaluing of this literature came from gothic fiction's "emerging readership of young women" (Kevin Corstorphine, *The Palgrave Handbook to Horror Literature* [Cham: Palgrave Macmillan, 2018), 1, 3]. Finally, at first glance, Petry's representation of the protagonist's wife, Lilly Belle, doesn't fit comfortably with readings of her work as feminist, and her characterization of the "Negro man" at the center "on Saturday" falls uncomfortably close to toxic white stereotypes of Black masculinity, similar to critiques of Richard Wright's exploration of Bigger Thomas in *Native Son* (1940). As I discuss later on in this chapter, Petry's use of aural imagery allows us deeper access to both of these characters, and the study of sound as a specialized artistic element of fiction writing has emerged only recently, enabling new understandings of canonical and so-called minor literary texts.

20  Claudia Rankine, *Just Us: An American Conversation* (Minneapolis, MN: Greywolf Press, 2020), 33. For research interrogating the sonic color line's racialization of space/spatialization of race in contemporary urban contexts see Allie Martin, "Hearing Change in the Chocolate City: Soundwalking as Black Feminist Method," *Sounding Out!* August 5, 2019, https://soundstudiesblog.com/2019/08/05/heari ng-change-in-the-chocolate-city-soundwalking-as-black-feminist-method/ ; Brandi Thompson Summers, "Reclaiming the Chocolate City: Soundscapes of Gentrification and Resistance in Washington D.C.," *Society and Space* (December 2020): 1–17; Tom Western, "Listening with Displacement: Sound, Citizenship, and Disruptive Representations of Migration," *Migration and Society* (2020): 194–309; and Pedro J. S. Viera de Oliveira, "Weaponizing Quietness: Sound Bombs and the Racialization of Noise," *Journal of the Design Studies Forum* (2019): 193–211, among others.

21  Paula M. L. Moya, *The Social Imperative: Race, Close Reading, and Contemporary Literary Criticism* (Stanford, CA: Stanford University Press, 2016), 35, 10.

22  Emily J. Lordi, *Black Resonance: Iconic Women Singers and African American Literature* (New Brunswick, NJ: Rutgers University Press, 2013), 183.

23  The trope of the listener is work I began in *The Sonic Color Line* (Stoever, *The Sonic Color Line: Race and the Cultural Politics of Listening* [New York: New

York University Press, 2016], 18–19). Although beyond the scope of this chapter, identifying Petry's use of aural imagery in "On Saturday" further solidifies her connection to earlier Black writers such as Frederick Douglass, a linkage that Kenneth Clark has made in his discussion of Petry's influences and "invocation of the gothic" and the "sequestration/confinement" of the moment where Douglass's enslaver forces him to listen as he beats Douglass's Aunt Hester (ibid., 21). The way in which Petry represents the siren as a source of nonverbal epistemology for both her protagonist and the reader is key to this connection— "sounds to be listened to for meaning, rather than irrational collateral noise" (ibid., 44)—as is the ending of the story where the siren transforms into "a louder sharper sobbing sound" and a "moan" reminiscent of a woman's scream, in Petry's narrator's case, one produced through his own hand (Petry, "On Saturday the Siren Sounds at Noon," 369). Such attentive analysis also reveals Petry's literary technique as a forerunner of, and precursor to, later representations of stratified aural imagery such as the prologue to Ralph Ellison's *Invisible Man*—where a similarly unnamed protagonist "slips into the breaks" of his "radio-phonograph" and "not only entered the music but *descended*, like Dante, into its depths" or strata! (Ralph Ellison, *Invisible Man* [New York: Vintage, 1972], 9, emphasis mine)—among other works that examine the layers of Black listening experience in relationship to sonic technologies: Du Bois's echo chamber in *Dusk of Dawn* (New Brunswick, NJ: Transaction, 1940) and Petry's radio in *The Street* (New York: Houghton Mifflin, 1946).
24  Michael Bull, *Sirens* (New York: Bloomsbury, 2020), 9.
25  See Rachel Poser's article on scholar Dan-el Padilla Peralta for a contemporary conversation about the role of "classics" in shaping white supremacy: "He Wants to Save Classics from Whiteness. Can the Field Survive?" *The New York Times*, February 2, 2021, https://www.nytimes.com/2021/02/02/magazine/classics-greece-rome-whiteness.html.
26  See Moya Bailey, "More on the origin of Misogynoir," *Moyazb*, April 27, 2014, https://moyazb.tumblr.com/post/84048113369/more-on-the-origin-of-misogynoir.
27  For a representation of an indigenous two-spirit person cursed as a siren, see: "A History of Violence," season 1, episode 4, *Lovecraft Country,* dir. Victoria Mahoney. Aired September 6, 2020, HBO.
28  Tananarive Due, "*Get Out* and the Black Horror Aesthetic," in Get Out: *The Complete Annotated Screenplay* (New York: Inventory Press, 2019), 7. Kenneth Clark's *The Radical Fiction of Ann Petry* builds a strong case for the revisioning of Petry as a writer of terror whose "imagination remained unbounded" (212). Evie Shockley has also identified *The Street* as a site of "Gothic homelessness" in "Buried Alive: Gothic Homelessness, Black Women's Sexuality, and (Living) Death in Ann Petry's *The Street*," *African American Review* 40, no. 3 (2006): 439–60.
29  James W. Ivey, "Ann Petry Talks About First Novel," *The Crisis* (January 1946): 48–9.
30  Jackson, *The Indignant Generation*, 144.
31  Petry, "On Saturday," 368.
32  Ibid.
33  This is a role Washington would assume at age 11, after his father passed. Unnamed in Petry's story but readily identifiable via the subway station's address, the name of the neighborhood—and family plantation—is Wakefield, built by and from the labor of the over 50 men and women enslaved by Washington's father. In 2021, Wakefield's

residents are primarily Black and Brown people of Caribbean ancestry; however, in the 1940s, the area was overwhelmingly white and of European descent, one of the Bronx's early "status neighborhoods" for German, Italian, and Irish immigrants who made enough money to leave the impoverished tenements of the South Bronx's "old neighborhoods." See Evelyn Gonzalez, *The Bronx* (New York: Columbia University Press, 2004), 102.

34   Federal Writer's Project, *The WPA Guide to New York City*, 1939 (New York: Pantheon Books, 1982), 540.

35   It also is reminiscent of one of Petry's literary influences, James Weldon Johnson's *The Autobiography of an Ex-Colored Man*, mentioned in Mark K. Wilson and Ann Petry, "A MELUS Interview: Ann Petry. The New England Connection," *MELUS* 15, no. 2 (1988): 74. Johnson's protagonist, who is passing as white, remains unnamed.

36   Petry, "On Saturday," 368.

37   Ibid.

38   Ibid.

39   Ibid.

40   This form of representation is called "subjective sound" in cinematic technique. In "Sound and Empathy: Subjectivity, Gender, and the Cinematic Soundscape," Robynn J. Stillwell describes that "point of audition" is slightly more common than true point of view camera work, but still used only intermittently by filmmakers. What is especially interesting to me is that she notes that in films where it is utilized—namely horror by Alfred Hitchcock—it has a dramatically different effect: "[W]hile point-of-view puts us in the subject position of a character in control, point-of-audition puts us in the subject position of a character who has lost or is losing control" (Robynn J. Stillwell, "Sound and Empathy: Subjectivity, Gender, and the Cinematic Soundscape," in *Film Music: Critical Approaches*, ed. Kevin J. Donnelly [New York: Continuum, 2001], 174).

41   Petry, "On Saturday," 369.

42   I have previously discussed Ann Petry's use of sound in *The Street* in *The Sonic Color Line*, 266–75.

43   Oxford English Dictionary Online, s.v. "stratified, adj.," accessed November 16, 2020, https://www.oed.com/.

44   Ellison, *Invisible Man*, 581.

45   While I can only gesture here, I'd love to think Petry's aural imagery in "On Saturday" and *The Street* with the emerging field of Black Time Studies, particularly via Julius B. Fleming, Jr.'s reading of Carter Mathes's *Imagine the Sound: Experimental African American Literature After Civil Rights* (Minneapolis: University of Minnesota Press, 2015) as exploring the "convergence of sound and black experimental writing . . . as a key window into comprehending the social, political, and aesthetic uses of time for freedom and subjugation, for tyranny and revolution" (Julius B. Fleming, Jr., "Sound, Aesthetics, and Black Time Studies," *College Literature* 46, no. 1 (2019): 282). Kristen L. Simmons and Kaya Naomi Williams's discussion of "the many and multiple presents of racial violence" is also extremely relevant to Petry's use of stratified aural imagery, in "I Was Dreaming When I Wrote This: A Mixtape for America," *Social Text* 36, no. 2 (2018): 146.

46   Raymond Daniell, "Coventry Dead in One Grave; Air Raid Siren Is their Requiem," *The New York Times*, November 21, 1940, https://timesmachine.nytimes.com/ti

mesmachine/1940/11/21/113117741.html?pageNumber=3. This article could very well have been source material for "On Saturday."

47 "Fire Siren Blast to Warn of Raids," *The New York Times*, December 9, 1941, 7.
48 "Urges a Cheerful Air Raid Siren," *The New York Times*, December 20, 1941.
49 "City Cocks Its Ears for Siren in Vain," *The New York Times*, December 18, 1941.
50 Ibid.
51 Ibid.
52 "New Air Raid Siren Does Best Job Yet," *The New York Times*, March 5, 1942.
53 Joseph Kinsley qtd. in "New Air Raid Siren to Be Tested Today," *The New York Times*, March 4, 1942. Demographic information from Gonzalez, *The Bronx*, 100.
54 "Listen in Vain for Air Raid Sirens," *The New York Amsterdam News*, December 20, 1942, 1.
55 "Air Raid Siren Test Set for Saturday," *The New York Times*, May 25, 1942.
56 "'Dead Spots' Are Reduced in Air Raid Siren Areas," *The New York Times*, July 14, 1942.
57 "Didn't Hear Sirens," *New York Amsterdam News*, August 15, 1942.
58 "New Air Raid Siren Does Best Job Yet," *The New York Times*, March 5, 1942.
59 Bull, *Sirens*, 37.
60 Ibid., 48, 79.
61 Ibid., 79.
62 See: Stoever, *The Sonic Color Line*.
63 "City Nonchalant as Sirens Wail," *The New York Times*, December 10, 1941.
64 Ibid. The piece also takes aim at a Harlem woman as having an excessive reaction, "immediately" quitting her job as a maid upon hearing the siren; it also employs sonic stereotypes to express antisemitism, claiming that a Jewish man in the Lower East Side got into an argument with an "Irish Cop" because he refused to stop selling his hot pretzels during the test.
65 Petry, "On Saturday," 368.
66 James G. Thompson, "Should I Sacrifice to Live Half-American?: Letter to the Editor," *The Pittsburgh Courier*, January 31, 1942.
67 Ruth Wilson Gilmore, *Golden Gulag: Prisons, Surplus, Crisis, and Opposition in Globalizing California* (Berkeley: University of California Press, 2007), 28.
68 Bull, *Sirens*, 37.
69 Petry, "On Saturday," 368.
70 Ann Petry, "Ann Petry," *Contemporary American Autobiography Series 6* (Detroit: Gale, 1988), 253–69. See also Carol Henderson, "The 'Walking Wounded': Rethinking Black Women's Identity in Ann Petry's *The Street*," *Modern Fiction Studies* 46, no. 4 (2000): 849–67.
71 Kevin L. Clay calls this "Black Resilience Neoliberalism" in "'Despite the Odds:' Unpacking the Politics of Black Resilience Neoliberalism," *American Educational Research Journal* 56, no. 1 (2019): 75–110. For a discussion of Black trauma and resilience in the contemporary context of COVID-19 and police brutality, see Chrystal Milner, "'It Just Weighs on Your Psyche': Black Americans on Mental Health, Trauma, and Resilience," *STAT*, July 6, 2020, https://www.statnews.com/2020/07/06/it-just-weighs-on-your-psyche-black-americans-on-mental-health-trauma-and-resilience/.
72 Petry, "On Saturday," 368.
73 Ibid.

74  Ibid., 368.
75  Rankine, *Just Us*, 31.
76  J. Martin Daughtry, *Listening to War: Sound, Music, Trauma, and Survival in Wartime Iraq* (Oxford: Oxford University Press, 2015), 98.
77  Bull, *Sirens*, 37.
78  Duke Ellington, *Music Is My Mistress* (New York: Doubleday, 1973), 103. "Lilly Belle May June" was also the title of a popular Django Reinhardt song released on Decca in 1935, with lyrics directing white supremacist longing for antebellum "Southern Belles" to a singer who has migrated North.
79  Dan Burley, "BACK DOOR STUFF: A 1940 Glamour Gal Wants To Retire To A Job," *New York Amsterdam News*, January 4, 1941, 16.
80  Petry, "On Saturday," 368.
81  For more on Du Bois and Black folk epistemologies in relation to sound, see Jennifer Stoever, "Fine-Tuning the Sonic Color-Line: Radio and the Acousmatic Du Bois," *Modernist Cultures* 10, no. 1 (2015): 99–118; Solimar Otero, *Archives of Conjure: Stories of the Dead in Afro-Latinx Cultures* (New York: Columbia University Press, 2020). "Mermaid" in Spanish is "la sirena."
82  Otero, *Archives of Conjure*, 6–7.
83  Ibid., 163–4.
84  Clark, *The Radical Fiction of Ann Petry*, 11.
85  Petry, "Ann Petry," 259 and 268. For more on the role of Black diasporic folk traditions and epistemologies in Petry's work see Clark, *The Radical Fiction of Ann Petry*, 19; Gladys J. Washington, "A World Cunningly Made: A Closer Look at Anne Petry's Short Fiction," in *Ann Petry's Short Fiction*, ed. Hazel Arnett Ervin and Hilary Holladay (Westport, CT: Praeger, 2004), 19; Melvin B. Rahming, "Phenomenology, Epistemology, Ontology, and Spirit: The Caribbean Perspective in Ann Petry's *Tituba of Salem Village*," *South Central Review* 20, no. 2/4 (2003): 24–46. She expressed this through references and plot points such as in *The Street*, when a character named Min visits a root doctor to obtain conjure items so that her boyfriend won't put her out on the street. In another little-known short story Petry wrote around the same time as "On Saturday" and bearing some similarity, one of the main characters is a Barbadian immigrant to New York City and dancer named Belle Rose—a pointed inversion of "Lilly Belle"—whose grandmother's knowledge as an "Obeah woman" enables her to use her "high shrill voice" to chant words the narrator "couldn't understand . . . the same kind of chant that a witch doctor uses when he casts a spell, the same one that the conjure women use and the obeah women" in order to prevent her long-lost boyfriend Olaf from killing her for her perceived transgressions as a night club dancer. The key stratified image in this story is the sound of the drums to which Belle Rose performs. Ann Petry, "Olaf and His Girl Friend," *The Crisis* (May 1945): 147.
86  Petry, "On Saturday," 369.
87  Petry remarked that she "had an absolute passion for motion pictures—I'm addicted to them: good ones, bad ones, and indifferent ones" and she had the experience of time dilation upon watching her first one, that "time should have come to a stop" while she was immersed in the movie's sight and sound, an experience echoed in the protagonist's incident of "On Saturday" ("Ann Petry," 267).
88  Petry, "On Saturday," 369.

89  W. E. B. Du Bois, *The Souls of Black Folk*, 1903 (New York: Penguin Books, 1989), 5.
90  Petry, "On Saturday," 368, 369.
91  Du Bois's notion of the veil—which I discuss shortly—draws on a Black Southern folk belief that infants who are born with a caul still attached, a portion of the birth membrane remaining on the head, are said to be 'behind the veil' and gifted with preternatural abilities, or "second sight." For more on Du Bois and Black folk epistemologies in relation to sound, see Stoever, "Fine-Tuning the Sonic Color Line," 99–118. See also Walter Rucker, "Conjure, Magic, and Power, The Influence of Afro-Atlantic Religious Practices on Slave Resistance and Rebellion," *Journal of Black Studies* 32, no. 1 (2001): 84–103, for a discussion of "how the near identical meaning of the caul among Africans in Dahomey, the Gold Coast, Dutch Guyana, Jamaica, Haiti, and the American South further demonstrates that enduring African spiritual concepts were lasting influences on the actions of enslaved Africans throughout the Atlantic world" (ibid., 94).
92  Petry, "On Saturday," 369.
93  Ibid., 368–9.
94  Ibid., 368.
95  Ibid., 369.
96  Ibid.
97  Stoever, *The Sonic Color Line*, 262–74.
98  Angela Y. Davis, *Blues Legacies and Black Feminism: Gertrude "Ma" Rainey, Bessie Smith, and Billie Holiday* (New York: Vintage, 1998), 12.
99  Ibid., 25.
100 Petry, "On Saturday," 369. Petry uses a similar image of "sheer, thin curtains blowing in the breeze" to spark a flashback of intimate partner violence in *The Street*, when Boots Smith remembers attempting to strangle his girlfriend Jubilee after he catches her cheating on him. Unlike Lilly Belle, Jubilee is able to fight back with a knife and escape the apartment alive. In an especially revealing moment, Petry characterizes Boots as taking pleasure in and power from the sound of Jubilee's screams because it meant "she was afraid of him because he was going to kill her, and he wanted her to know it beforehand and be afraid." He continues to take power from this sound as she flees, realizing "with all that noise and screaming no one had tried to find out what was going on. He could have killed her easy and no one would even have rapped on the door" (Petry, *The Street*, 270). In a particular kind of intersectional violence, American society mutes Black women's screams of pain of meaning and urgency, silencing them as everyday background noise marking the collateral damage of the entangled ideologies of white supremacy and toxic patriarchal masculinity.
101 Du Bois, *The Souls of Black Folk*, 5.
102 Saidiya Hartman, *Wayward Lives, Beautiful Experiments: Intimate Histories of Social Upheaval* (New York: W.W. Norton, 2019), 166.
103 Ibid., 336.
104 Petry, "On Saturday," 369.
105 Holladay, *Ann Petry*, 11.
106 Petry, "On Saturday," 369.
107 Tsitsi Jaji, *Africa in Stereo: Modernism, Music, and Pan-African Solidarity* (Oxford University Press, 2014), 173.

108   Ibid., 148.
109   Mathes, *Imagine the Sound*, 195.
110   The Digital Scholarship Lab and the National Community Reinvestment Coalition, "Not Even Past: Social Vulnerability and the Legacy of Redlining," in *American Panorama*, ed. Robert K. Nelson and Edward L. Ayers, 2020. https://dsl.richmond.edu/socialvulnerability. Other US cities are searchable here too.
111   Dr. Moore publicly attested to white doctors' refusal to listen to her complaints of pain—"they made me feel like a drug addict"—and to the inadequacy of her colleagues' care—"he did not even listen to my lungs, he didn't touch me in any way"—even as she knowledgeably requested proper treatment. See: Aletha Maybank, Camara Phyllis Jones, Uché Blackstock, and Joia Crear Perry, "Say Her Name: Dr. Susan Moore," *The Washington Post*, December 26, 2020, https://www.washingtonpost.com/opinions/2020/12/26/say-her-name-dr-susan-moore/.

## Works Cited

Associated Press. "Birdsong, Blaring Sirens: A Pandemic in Sound." Released April 6, 2020. Video, 1:13. https://www.youtube.com/watch?v=UlNVa5qME64.

Associated Press. "Hopeful Birdsong, Foreboding Sirens: A Pandemic in Sound." *Associated Press News*, April 11, 2020. https://apnews.com/article/d18d4fc6e2f4ff3f1d439fc12f6a3d98.

Bachom, Sandi. "DAY28: Sirens and Refrigerated Trucks At NYC Hospitals 3/29/20." Released March 30, 2020. Video, 4:56. https://www.youtube.com/watch?v=4WmJ-f3SB4c.

Bailey, Moya. "More on the Origin of Misogynoir." *Moyazb*, April 27, 2014. https://moyazb.tumblr.com/post/84048113369/more-on-the-origin-of-misogynoir.

Brooks, Paula L. *Spooks, Spies, and Private Eyes: Black Mystery, Crime and Suspense Fiction of the 20th Century*. New York: Doubleday, 1995.

Bui, Quoctrung, and Emily Badger, "The Coronavirus Quieted City Noise: Listen to What's Left." *The New York Times*, May 22, 2020. https://www.nytimes.com/interactive/2020/05/22/upshot/coronavirus-quiet-city-noise.html.

Bull, Michael. *Sirens*. New York: Bloomsbury, 2020.

Burley, Dan. "BACK DOOR STUFF: A 1940 Glamour Gal Wants To Retire To A Job." *New York Amsterdam News*, January 4, 1941.

Clark, Kenneth. *The Radical Fiction of Ann Petry*. Baton Rouge: Louisiana State University Press, 2013.

Clay, Kevin L. "'Despite the Odds:' Unpacking the Politics of Black Resilience Neoliberalism." *American Educational Research Journal* 56, no. 1 (2019): 75–110.

Corstorphine, Kevin. *The Palgrave Handbook to Horror Literature*. Cham: Palgrave Macmillan, 2018.

Daniell, Raymond. "Coventry Dead in One Grave; Air Raid Siren is their Requiem." *The New York Times*, November 21, 1940. https://timesmachine.nytimes.com/timesmachine/1940/11/21/113117741.html?pageNumber=3.

Daughtry, J. Martin. *Listening to War: Sound, Music, Trauma, and Survival in Wartime Iraq*. Oxford: Oxford University Press, 2015.

Davis, Angela Y. *Blues Legacies and Black Feminism: Gertrude "Ma" Rainey, Bessie Smith, and Billie Holiday*. New York: Vintage, 1998.
de Luca, Antonio, Dave Taft, and Umi Syam. "New York Is Quiet: Listen to the Birds," *The New York Times*, May 31, 2020. https://www.nytimes.com/interactive/2020/05/31/nyregion/coronavirus-birding-nyc.html.
The Digital Scholarship Lab and the National Community Reinvestment Coalition. "Not Even Past: Social Vulnerability and the Legacy of Redlining." In *American Panorama*, edited by Robert K. Nelson and Edward L. Ayers, 2020. https://dsl.richmond.edu/socialvulnerability.
Du Bois, W. E. B. *The Souls of Black Folk*. 1903. New York: Penguin Books, 1989.
Du Bois, W. E. B. *Dusk of Dawn*. 1940. New Brunswick: Transaction Publishers, 2005.
Due, Tananarive. "*Get Out* and the Black Horror Aesthetic." In Get Out: *The Complete Annotated Screenplay*, 6–14. New York: Inventory Press, 2019.
Ellington, Duke. *Music Is My Mistress*. New York: Doubleday, 1973.
Ellison, Ralph. *Invisible Man*. 1952. New York: Vintage, 1972.
Emmer, Pascal, Woods Ervin, Derecka Purnell, Andrea J. Ritchie, and Tiffany Wang. *Unmasked: Impacts of Pandemic Policing*. COVID19 Policing Project, 2020.
Ervin, Hazel Arnett, and Hillary Holladay, eds. *Ann Petry's Short Fiction: Critical Essays*. Westport, CT: Praeger, 2004.
Federal Writer's Project. *The WPA Guide to New York City*. 1939. New York: Pantheon Books, 1982.
Fleming, Julius B. Jr. "Sound, Aesthetics, and Black Time Studies." *College Literature* 46, no. 1 (2019): 282.
Giovetti, Olivia. "Notes on Birdsong: A Dual History of Racism and Nature's Music." *Van Magazine*, May 29, 2020. https://van-us.atavist.com/notes-on-birdsong.
Gonzalez, Evelyn. *The Bronx*. New York: Columbia University Press, 2004.
Hartman, Saidiya. *Wayward Lives, Beautiful Experiments: Intimate Histories of Social Upheaval*. New York: W.W. Norton, 2019.
Holladay, Hillary. *Ann Petry*. New York: Twayne Publishers, 1996.
Ivey, James W. "Ann Petry Talks About First Novel." *The Crisis* (January 1946): 48–9.
Jackson, Lawrence P. *The Indignant Generation: A Narrative History of African American Writers and Critics, 1934–1960*. Princeton, NJ: Princeton University Press, 2011.
Jaji, Tsitsi. *Africa in Stereo: Modernism, Music, and Pan-African Solidarity*. Oxford: Oxford University Press, 2014.
Jong-Fast, Molly. "Embracing Life in a Time of Death." *Vogue*, April 18, 2020. https://www.vogue.com/article/new-york-coronavirus-life-in-time-of-death.
Kalaf, Samer. "Alone in the City of Sirens." *New Republic*, April 10, 2020. https://newrepublic.com/article/157256/coronavirus-new-york-city-sirens.
Krenz, Karissa. "How the Sound of New York City has Changed during the COVID-19 Lockdown." *WQXR Editorial*, April 30, 2020. https://www.wqxr.org/story/how-sound-new-york-city-has-changed-during-covid-19-lockdown/.
Lewak, Doree. "New Yorkers are on Edge Over Relentless Sirens During Coronavirus Lockdowns." *New York Post*, April 1, 2020. https://nypost.com/2020/04/01/new-yorkers-on-edge-over-relentless-sirens-in-coronavirus-lockdown/.
Lordi, Emily J. *Black Resonance: Iconic Women Singers and African American Literature*. New Brunswick, NJ: Rutgers University Press, 2013.
Mahoney, Victoria, dir. *Lovecraft Country*. Season 1, episode 4, "A History of Violence." Aired September 6, 2020, HBO.

Martin, Allie. "Hearing Change in the Chocolate City: Soundwalking as Black Feminist Method." *Sounding Out!* August 5, 2019. https://soundstudiesblog.com/2019/08/05/hearing-change-in-the-chocolate-city-soundwalking-as-black-feminist-method/.

Mathes, Carter. *Imagine the Sound: Experimental African American Literature After Civil Rights.* Minneapolis: University of Minnesota Press, 2015.

Maybank, Aletha, Camara Phyllis Jones, Uché Blackstock, and Joia Crear Perry. "Say Her Name: Dr. Susan Moore." *The Washington Post*, December 26, 2020. https://www.washingtonpost.com/opinions/2020/12/26/say-her-name-dr-susan-moore/.

Milner, Chrystal. "'It just weighs on your psyche': Black Americans on Mental Health, Trauma, and Resilience." *STAT*, July 6, 2020. https://www.statnews.com/2020/07/06/it-just-weighs-on-your-psyche-black-americans-on-mental-health-trauma-and-resilience/.

Moya, Paula M. L. *The Social Imperative: Race, Close Reading, and Contemporary Literary Criticism.* Stanford: Stanford University Press, 2016.

*The New York Amsterdam News*. "Didn't Hear Sirens." August 15, 1942.

*The New York Amsterdam News*. "Listen in Vain for Air Raid Sirens." December 20, 1942.

New York State Government. "Governor Cuomo Signs the 'New York State on PAUSE' Executive Order." March 20, 2020. https://www.governor.ny.gov/news/governor-cuomo-signs-new-york-state-pause-executive-order.

New York State Government. "Amid Ongoing COVID-19 Pandemic, Governor Cuomo Announces New Hospital Network Central Coordinating Team." March 31, 2020. https://www.governor.ny.gov/news/amid-ongoing-covid-19-pandemic-governor-cuomo-announces-new-hospital-network-central.

*The New York Times*. "Air Raid Siren Test Set for Saturday." May 25, 1942.

*The New York Times*. "City Cocks Its Ears For Siren In Vain." December 18, 1941.

*The New York Times*. "City Nonchalant as Sirens Wail." December 10, 1941.

*The New York Times*. "Coronavirus World Map: Tracking the Outbreak." January 19, 2021. https://www.nytimes.com/interactive/2020/world/coronavirus-maps.html.

*The New York Times*. "'Dead Spots' Are Reduced in Air Raid Siren Areas." July 14, 1942.

*The New York Times*. "Fire Siren Blast to Warn of Raids." December 9, 1941.

*The New York Times*. "New Air Raid Siren Does Best Job Yet." March 5, 1942.

*The New York Times*. "New Air Raid Siren To Be Tested Today." March 4, 1942.

*The New York Times*. "Urges a Cheerful Air Raid Siren." December 20, 1941.

Otero, Solimar. *Archives of Conjure: Stories of the Dead in Afro-Latinx Cultures.* New York: Columbia University Press, 2020.

Oxford English Dictionary Online. s.v. "stratified, adj." accessed November 16, 2020. https://www.oed.com/.

Peele, Jordan, dir. *Get Out.* 2017; Universal Pictures. Film.

Petry, Ann. "On Saturday the Siren Sounds at Noon." *The Crisis* 50, no. 12 (December 1943): 368–9.

Petry, Ann. "Olaf and His Girl Friend." *The Crisis* 52, no. 5 (May 1945): 135–8, 147.

Petry, Ann. *The Street.* New York: Houghton Mifflin, 1946.

Petry, Ann. "Ann Petry." *Contemporary American Autobiography Series* 6, 253–6. Detroit: Gale, 1988.

Poser, Rachel. "He Wants to Save Classics From Whiteness: Can the Field Survive?" *The New York Times*, February 2, 2021. https://www.nytimes.com/2021/02/02/magazine/classics-greece-rome-whiteness.html.

Rahming, Melvin B. "Phenomenology, Epistemology, Ontology, and Spirit: The Caribbean Perspective in Ann Petry's *Tituba of Salem Village*." *South Central Review* 20, no. 2/4 (2003): 24–46.

Rankine, Claudia. *Just Us: An American Conversation*. Minneapolis: Greywolf Press, 2020.

Remnick, David. "New York City in the Coronavirus Pandemic." *The New Yorker*, April 5, 2020. https://www.newyorker.com/magazine/2020/04/13/new-york-city-in-the-coronavirus-pandemic.

Rucker, Walter. "Conjure, Magic, and Power, The Influence of Afro-Atlantic Religious Practices on Slave Resistance and Rebellion." *Journal of Black Studies* 32, no. 1 (2001): 84–103.

Shockley, Evie. "Buried Alive: Gothic Homelessness, Black Women's Sexuality, and (Living) Death in Ann Petry's *The Street*." *African American Review* 40, no. 3 (2006): 439–60.

Simmons, Kristen L., and Kaya Naomi Williams. "I Was Dreaming When I Wrote This: A Mixtape for America." *Social Text* 36, no. 2 (2018): 145–64.

Sims, John. *2020: (Di)Visions of America*. The Ringling Museum (Sarasota, Florida), January 18, 2021. Virtual streaming program. https://www.ringling.org/2020-divisions-america-performance-john-sims.

*State of Black America, Unmasked 2020: Executive Summary*. A National Urban League Publication, 2020. http://sobadev.iamempowered.com/sites/soba.iamempowered.com/files/NUL-SOBA-2020-ES-web.pdf.

Stillwell, Robynn J. "Sound and Empathy: Subjectivity, Gender, and the Cinematic Soundscape." In *Film Music: Critical Approaches*, edited by Kevin J. Donnelly, 167–87. New York: Continuum, 2001.

Stoever, Jennifer Lynn. "Fine-tuning the Sonic Color-line: Radio and the Acousmatic Du Bois." *Modernist Cultures* 10, no. 1 (2015): 99–118.

Stoever, Jennifer Lynn. *The Sonic Color Line: Race and the Cultural Politics of Listening*. New York: New York University Press, 2016.

Stokstad, Erik. "When COVID-19 Silenced Cities, Birdsong Recaptured Its Former Glory." *Sciencemag*, September 24, 2020. https://www.sciencemag.org/news/2020/09/when-covid-19-silenced-cities-birdsong-recaptured-its-former-glory.

Summers, Brandi Thompson. "Reclaiming the Chocolate City: Soundscapes of Gentrification and Resistance in Washington D.C." *Society and Space* 39, no. 1 (December 2020): 1–17.

Thompson, James G. "Should I Sacrifice to Live Half-American?: Letter to the Editor." *The Pittsburgh Courier*, January 31, 1942.

Vera, Amir, and Laura Ly. "White Woman Who Called Police on a Black Man Bird-Watching in Central Park has been Fired." *CNN*, May 26, 2020. https://www.cnn.com/2020/05/26/us/central-park-video-dog-video-african-american-trnd/index.html.

Viera de Oliveira, Pedro J. S. "Weaponizing Quietness: Sound Bombs and the Racialization of Noise." *Journal of the Design Studies Forum* 11, no. 2 (2019): 193–211.

Vitullio-Martin, Julia. "Sirens and Suffering: Rethinking the Soundtrack of the Coronavirus Crisis." *Gotham Gazette*, April 21, 2020. https://www.gothamgazette.com/opinion/9324-sirens-suffering-rethinking-soundtrack-coronavirus-crisis-new-york-city.

Washington, Gladys J. "A World Cunningly Made: A Closer Look at Anne Petry's Short Fiction." In *Ann Petry's Short Fiction*, edited by Hazel Arnett Ervin and Hilary Holladay, 1–12. Westport, CT: Praeger, 2004.

Western, Tom. "Listening with Displacement: Sound, Citizenship, and Disruptive Representations of Migration." *Migration and Society* 3 (2020): 194–309.

Wilson, Mark K., and Ann Petry. "A MELUS Interview: Ann Petry. The New England Connection." *MELUS* 15, no. 2 (1988): 71–84.

Wilson Gilmore, Ruth. *Golden Gulag: Prisons, Surplus, Crisis, and Opposition in Globalizing California.* Berkeley: University of California Press, 2007.

Zoladz, Lindsey. "Learning to Listen to, and Beyond, the Siren's Call." *The New York Times*, April 5, 2020. https://www.nytimes.com/2020/04/05/arts/music/coronavirus-sirens-music.html.

# Contributors

**Nathalie Aghoro** is Assistant Professor of North American Literary and Cultural Studies at the Catholic University of Eichstätt-Ingolstadt, Germany. Her research interests include auditory culture, postmodern and contemporary literature, media studies, as well as social justice and public sphere studies. She is the author of *Sounding the Novel: Voice in Twenty-First Century American Fiction* (2018), the co-editor of the 2017 JCDE special issue on *Theatre and Mobility* (with Kerstin Schmidt), and an editorial board member of the De Gruyter Video Games and the Humanities series. Her current book project deals with the ties between social justice, solidarity, and cultural practices in shared places.

**Nicole Brittingham Furlonge** is Klingenstein Family Chair Professor and Director of the Klingenstein Center, Columbia University, USA. She is the author of *Race Sounds: The Art of Listening in African American Literature* (2018) and has published in journals such as *Callaloo*, *Interference*, and *Sounding Out!* She also consults with and teaches seminars on listening and race in the Narrative Medicine program at Columbia University Vagelos College of Physicians and Surgeons. Her LEARNS Collaborative works to catalyze human-centered, equitable change in schools, organizations, and communities by helping people listen differently to themselves and each other. Her research and teaching interests focus on the intersections between listening, cognitive neuroscience, social justice, and leadership. Currently, she is working on a sonic cultural history of Webster Hall in New York City and, through a cross-Atlantic collaboration with Salomé Voegelin and Mark Wright of Listening Across Disciplines II, an interdisciplinary and dynamic exploration of Sonic Pedagogies.

**Christof Decker** is Professor of American Studies at the Ludwig-Maximilians-Universität Munich, Germany. His research encompasses film, media, and visual culture studies, nineteenth- and twentieth-century American literature and culture, and media aesthetics. His most recent book publication is *Transnational Mediations: Negotiating Popular Culture between Europe and the United States* (2015, co-edited with Astrid Böger). Recent essays have considered Ben Shahn and the art of the war poster as well as love and politics in the cinema of the 1930s and 1940s.

**Julius Greve** is Lecturer and Research Associate at the Institute for English and American Studies, University of Oldenburg, Germany. He is the author of *Shreds of Matter: Cormac McCarthy and the Concept of Nature* (2018), and of numerous essays on McCarthy, Mark Z. Danielewski, François Laruelle, and critical theory. Greve has co-edited *America and the Musical Unconscious* (2015), *Superpositions: Laruelle and the Humanities* (2017), "Cormac McCarthy between Worlds" (*EJAS*-special issue,

2017), *Spaces and Fictions of the Weird and the Fantastic: Ecologies, Geographies, Oddities* (2019), and *The American Weird: Concept and Medium* (2020). He is currently working on a manuscript that delineates the relation between modern poetics and ventriloquism.

**Florian Groß** teaches American Studies at the Leibniz University Hannover, Germany. He is currently finishing his PhD thesis "Negotiating Creativity in Post-Network Television Series." Next to American television culture, his research interests include comics and graphic novels, contemporary literature, questions of authenticity in relation to contemporary notions of creativity, and the cultural history of New York City. He is co-editor of *The Aesthetics of Authenticity: Medial Constructions of the Real* (2012) and has published articles on the television series *30 Rock*, Michael Chabon's novel *The Amazing Adventures of Kavalier & Clay*, and the High Line Park as well as world's fairs in New York City.

**Tsitsi Jaji** is Associate Professor of English, African, and African American Studies at Duke University, USA. She is the author of *Africa in Stereo: Music, Modernism and Pan-African Solidarity* (2014). The book won the African Literature Association's First Book Prize and honorable mentions from the American Comparative Literature Association and Society for Ethnomusicology. She is also a poet and the author of *Mother Tongues* (2018), *Beating the Graves* (2017), and the chapbook *Carnaval* (2014). Currently she is working on two research projects: *Cassava Westerns*, a study of how global Black writers and artists reimagine the American frontier myth to serve new, local purposes; and *Classic Black*, a study of poetry set to music by Black concert music composers.

**Sabine Kim** is a postdoc researcher affiliated with the Obama Institute for Transnational American Studies at Mainz University, Germany. Her publications include the monograph *Acoustic Entanglements: Sound and Aesthetic Practice* (2017) as well as articles on the transatlantic travels of nineteenth-century African Americans, and Indigenous challenges to colonial dispossession. Her current research project looks at the relationship between wealth and waste. She is the managing editor of the *Journal of Transnational American Studies*.

**Irene Polimante** has recently earned her PhD in Modern Languages and Literature at the University of Macerata, Italy. Her dissertation deals with performance poetry as a tactile-kinesthetic *poiesis*. Her studies focus on the influences of performance poetry on the contemporary American artistic and transcultural scenario, thanks to an interdisciplinary and intersectional approach. On the topic she has published articles and essays like "Ameriscopia" (2017), "Tracie Morris's Poetic Experience" (2018), "Performing Identity" (2019), and "A Bluesy Sound" (2020). Her research interests include critical theory, postcolonial studies, cultural studies, media, and sound studies. She is currently working on contemporary hybrid forms of communicative, poetic, and performative art, which address the paradox of intimacy and distance in today's hyperconnected reality.

**Jennifer Lynn Stoever** is Associate Professor at Binghamton University, USA. She teaches courses on African American literature, sound studies, and race and gender representation in popular music. She is the author of *The Sonic Color Line: Race and the Cultural Politics of Listening* (2016) and has published in journals such as *Social Text, Social Identities, Sound Effects, American Quarterly, Radical History Review,* and *Modernist Cultures*. She is also the Co-Founder and Editor-in-Chief of *Sounding Out!: The Sound Studies Blog* and the project coordinator for the Binghamton Historical Soundwalk Project—a multiyear archival, civically engaged art project designed to challenge how Binghamton students and year-round residents hear their town, themselves, and each other.

**Allison Whitney** is Associate Professor of Film and Media Studies in the Department of English at Texas Tech University, USA. She has published on cinephilia, film technology, and sonic literacy in publications like *For the Love of Cinema* (2017), *Star Wars and the History of Transmedia Storytelling* (2017), and *Music, Sound and The Moving Image*. She is the co-editor of "Re-Thinking the Film History Survey" in the *Journal of Cinema and Media Studies Teaching Dossier* (with Paul McEwan, 2019). Currently, she is working on the representation of space exploration in cinema, oral histories of film exhibition culture, and a book project on IMAX film.

# Index

acousmêtre 119
acoustic cues 124
acoustic palimpsests 16, 20
acoustic properties 115, 122, 124
acoustics 122–7
Adorno, Theodor W. 47, 48
*Africa in Stereo* 181
*agent provocateur* 29
Album as Counter-Archive 83–5
Allen, Danielle 23
alternative culture 130, 131, 133, 136, 137, 140, 141
alternative music 131, 132, 134, 135, 138, 139
  culture 131, 135
Altman, Robert 44, 45, 49–53
America in a Cycle of Sonnets 32–4
*American Sonnets for My Past and Future Assassin* 32
Anderson, Benedict 2
articulation 63–6, 71
"Articulation of Sound Forms in Time" 61
Aslinger, Ben 132, 136
Atherton, Reverend Hope 62
*The Auditory Culture Reader* 1
auditory style 53
aural dimension 29
aural imagery 166–8, 170, 177, 182
autonomous music 48

Back, Les 1
Baker, Theodore 89
Bakhtin, Michael 29
Baldwin, James 33
*The Battle of Little Big Horn* 99, 100
Baudrillard, Jean 65
Beck, Jay 51
Beck, Les 48
Belle, Lilly 168, 177–81
Belmore, Rebecca 90

Berlant, Lauren 3
Berlitz 35
Bickford, Susan 23
Black workers 116–19
*Blues Legacies and Black Feminism* 180
Boas, Franz 84
bodies 37
  of meaning 29
Bordwell, David 49
Brathwaite, Edward Kamau 108
*Broken Arrow* 117
*Broken Calypsonian* 30
Brooks, Daphne 13
Browner, Tara 88
Buhler, James 46, 47, 49
Bull, Michael 1, 48
Burnetts, Charles 121
Buzzworm 147–51, 155–8

*Californication* 134, 140
*The Call* 115–28
Canadian Museum of History 84
capitalism 132–4
ceremonies 78, 81–3, 88, 91
children 18, 22, 62–4, 80, 118, 123, 168, 170, 176, 177, 179
Chion, Michel 119
cinema 48–50, 119
*City of Quartz* 149
Civil, Gabrielle 29, 34–6
Clarke, George Elliott 68
*A Clockwork Orange* 50
Clover, Carol 120
Coleman, Robin Means 34, 121
colonialism 78, 79, 86, 88, 100–3, 108
*Composing for the Films* 47
*The Conquest of Cool* 132
consumption 131–3, 138, 139
*The Conversation* 50
Cooke, Mervyn 46
cool 18, 130–4, 136, 137, 139–41

*Corregidora* 20–3
Corregidora women 20–3
Coulthard, Glen 88
counterculture 50, 132, 133, 140, 141
Coursil, Jacques 101–3, 107
  music 103, 104
COVID-19 7, 164–8, 182
creativity 132–4
crime film 115
critical theories 46, 47, 49
cross-media approach 4
cultural politics, of refusal 78–96
cultural sites 2

Daughtry, J. Martin 16, 20, 176
Davis, Angela Y. 180
Davis, Mike 149
Davis, Miles 15, 16
dead spots 164–91
*Debths* 71
deep listening 48
Deleuze, Gilles 65
democratic voice 51
Derrida, Jacques 73 n.5
Dickinson, Emily 33
*Discrepant Engagement: Dissonance, Cross-Culturality, and Experimental Writing* 61
disembodied voice 52, 116–19, 181
*Djbot Baghostus's Run* 15
double-consciousness 178
Du Bois, W. E. B. 176, 179
DuPlessis, Rachel Blau 63
Dutcher, Jeremy 78–96

*Easy Rider* 50
Ebner, Shannon 71
echo 61, 62, 64, 66, 71, 107, 165, 170, 172, 175, 176, 181, 182
echological environments 67–71
echology 61, 62, 64, 66, 68, 69, 71
ecology 61, 62, 66, 71
Eisler, Hanns 47, 48
Ellington, Duke 177
Embodiment 67, 68, 71, 116, 119
Enduring Women 115, 120–1, 125, 127
*Experiments in Joy* 35

"Fascinated to Presume: In Defense of Fiction" 17
film aesthetics 44–56
film music 45, 46, 48, 49, 53
film sound 45, 119
Final Girls 115, 117, 119–27
*Five Easy Pieces* 50
Frank, Thomas 132
Frederick, Rico 29–32

Gilmore, Ruth Wilson 175
Gilroy, Paul 9 n.22
Gitelman, Lisa 90
Glissant, Edouard 102, 103, 107
Guattari, Félix 65
Gunning, Tom 118

Hanson, Helen 45
Hayes, Terrance 29, 32–4
*Hearing the Movies* 46
*HeartBreaker* 31
Heath, Joseph 133
Heidi Kiiwetinepinesiik Stark 78
Hendrix, Jimi 33
Hilmes, Michelle 156
*History of the Voice* 108
horror 18, 20, 115, 120, 121, 124, 125, 168–70
horror films 115, 120, 121, 124
*Horror Noire: Blacks in American Horror Films from the 1890s to the Present* 121
hot spots 164–91
Howe, Susan 61–8, 71
*The Hunt for Red October* 117

imagined political community 2
indie rock 137
Indigenous music 84, 88, 89
Indigenous peoples 78, 82, 84–6, 88, 91, 100, 104
Indigenous sonic archives 84
  reframing 78–96
interior acoustics 21
*In the Wake: On Blackness and Being* 67
*The Invention of Creativity* 133
invisible labor 116–19

James, Robin  3
*Jaws*  49
jazz  16–19, 102–8
*Jazz*  17–19
Johnson, Gaye Theresa  154
Jones, Gayl  20
Jong-Fast, Molly  164
*Just Us*  166

Kara, Esen  153
Keeling, Kara  2
Kevorkian, Martin  116
Kun, Josh  2

LaBelle, Brandon  1, 2, 157
*Le Sentier des larmes : le grand exil des Indiens Cherokees*  106
listening  14–21, 23–4, 48, 84, 99, 103, 109, 147, 152, 155–6, 167
   practices  14, 17, 23, 158
   strain of  108–9
listening in print  13–26
   as discrepant practice  14–17
   pause  23–4
   trouble in mind and inner listening  20–3
   word and the sound  17–20
literacy 15, 30
   acoustic 2, 4
   sonic 115, 117, 121, 124
lived experience, dis-embodiment  71
logos  14
*The Long Goodbye*  50
Lordi, Emily  167

MacDowell, Edward  89
McGilligan, Patrick  51
Mackey, Nathaniel  14, 61, 71
*Mad Men*  134, 135, 140–1
mainstream  50, 51, 80, 131–3, 136–42
maps  66, 106, 107, 118, 148–50, 153–6
Martin, Michèle  117
*MASH*  51
Mathes, Carter  182
Mechling, William H.  84
memories  21–3, 30, 33, 34, 79, 86, 169, 170, 178, 179
Middle Passage  69, 104–6

*Minimal Brass*  103
mirrored lament  104–7
Monkman, Kent  89
monolithic entity  15
Montgomery, Will  67
Morrison, Toni  13, 15, 17–20, 33
Morse code  30
museums  79–84, 91
music  13, 16, 18, 31, 32, 46, 48, 49, 89, 107, 122, 124, 131, 132, 134–8, 156
musical performances  52, 152
musical taste  135–7
musical work  100, 101, 149
music industry  32, 52, 133
music percussion  18

narrative voice  150, 151
narrator  17, 19, 20, 167, 178, 179
*Nashville*  44, 45, 49–53
Native American  62–6, 89, 91, 99, 101, 104, 106
neoliberalism  3
Neumeyer, David  46
New England  61, 66, 67, 167
New Hollywood Cinema  44–56
Nichols, Bill  45
noise  28, 46, 49, 50, 52, 90, 147, 152–4, 157, 174, 178

*The O.C.* effect  134–41
"On Saturday the Siren Sounds at Noon"  164–91
opacity  50, 103, 107, 180
oppositional aesthetic movement  50
orchestral maps  152–4
overpass  150, 154, 157
*The Oxford Handbook of Film Music Studies*  46

pandemics  165, 166, 182
Patsavas, Alexandra  135, 141
Pearl Harbor  171, 172
performance  15, 17, 28, 29, 31, 34–7, 79, 80, 84, 87–9
Perloff, Marjorie  64
Petry, Ann  7, 164–91
Philip, M. NourbeSe  61–2, 67–9

# Index

phonographies 3
place 16–19, 46, 65, 108, 148–52, 154, 156, 158
   politics 148–52
poetics 16, 29, 32–4, 61, 62, 68, 69, 71, 151
poetic trans-positions 30–2
poetry 28–31, 34–7, 64, 67, 68, 71
political ventriloquism 68
politics of recognition 88–91
politics of sound 7
Pooch, Melanie 151
Potter, Andrew 133
power 122–7

Quashie, Kevin 23

*Race Sounds* 14, 15
racism 175
Rankine, Claudia 166
*The Rebel Sell* 133
Reckwitz, Andreas 133
recorded music 15, 16
recorded sound, refusal and temporalities 85–8
recordings 44, 45, 51, 79, 82, 85, 86, 90, 106, 124, 158
*Red Skin, White Masks: Rejecting the Colonial Politics of Recognition* 88
regionalist acoustics 62–7

Scarry, Elaine 15
Schatz, Thomas 49
sculpture 101
serial television, alternative soundtrack 134–9
settler colonialism 78, 79, 100, 108
*Seven Steps to Heaven* 15, 16
Sharpe, Christina 67
*She Tries Her Tongue, Her Silence Softly Breaks* 62, 67
silent period 45
simultaneous sounds 167, 171
*Singularities* 62, 64, 65, 67
sirens 7, 164–7, 169, 170, 172–8, 182
siren sounds 168–70

slavery 20, 21, 36, 70, 100, 103, 106
Smith, Bruce R. 15
Smith, Zadie 17
social 14, 34, 66, 68, 70, 84, 85, 87, 118, 134, 147, 148, 150–4, 156–8; *see also individual entries*
social change 1
   and sonic subversion 154–7
social dynamics 147, 152
social imaginaries 148–52, 157
social interactions 149, 150
sociality 87
social relations 66, 152, 156, 157
social space 61, 154
social worlds 87, 165
sonic color line 7 n.2, 84, 166, 167, 173–5, 180–2
sonic episteme 3
sonic fingerprint 136
sonic imaginaries 2
sonic repatriation 80–3
sonic sites, subversion 147–62
sonic subversion 148, 152, 154, 157, 158
   politics 148–52
   potentials and limitations of 157–8
   and social change 154–7
Sontag, Susan 49
*The Souls of Black Folk* 180
soundmapping 149, 152, 153, 155, 156
sounds 16, 17, 19, 28, 31, 37, 44–56, 101, 102, 105, 124, 148, 150, 151, 157, 165
   collecting 88–91
   effects 31, 46, 51
   in Film Studies 45–8
   forms 61–3, 70
   poetry 28
   practice 1-4
   recordings 51, 80, 83, 90, 148, 155
   studies 14, 45, 48, 61, 71, 102, 166
   style(s) of 48–9
   technology 45
soundscapes 7 n.1, 53, 87, 137, 138, 150, 152, 155, 166, 173
soundtrack 46–8, 50, 119, 125, 134, 136–8, 140, 141
Sow, Ousmane Huchard 99

space 122–7
Spielberg, Steven 49
Sterne, Jonathan 45
Stillwell, Robynn J. 187 n.40
Stoever, Jennifer Lynn 84
storyteller 15, 19, 20
stratified aural imagery 170–1
subculture 49

*Talking to Strangers* 23
Taylor, Charles 2
*Teaching the Lost* 89
Telephone Operator 115, 117
television series 131, 135, 138
"Thorow" 61
*A Thousand Plateaus* 65
Tiomkin, Dimitri 47
*Trails of Tears* 101–3
transient sounds 152–4
Trans-media creative flow 34–6
*Tropic of Orange* 147–62
tune, changing the 154–7
Tyler, Stephen 32

unmixing, practices 44–56
unsettled scores 99–111

urban soundscapes 31, 157
Ursa's voice 21

ventriloquism 67–70
Vincent, Bernard 106
visual style 52, 53
Voegelin, Salomé 8 n.19
voice 21, 28, 31, 51, 52, 61, 67, 79, 116, 117, 119, 123, 125–7, 151, 178, 179, 181
"The Voice of Documentary" 45

Walker, Hal Phillip 52
"Weavers of Speech: Telephone Operators as Defiant Domestics in American Culture and Literature" 116
Webb, Jim 51
Weheliye, Alexander 3
Wolastoqey language 93 n.8
*Wolastoqiyik Lintuwakonawa* 78–80
Woods, Faye 137

Yamashita, Karen Tei 7, 87, 147–62

*Zong!* 62, 70

www.ingramcontent.com/pod-product-compliance
Lightning Source LLC
Chambersburg PA
CBHW061828300426
44115CB00013B/2295